third edition

The Stages of Human Evolution

HUMAN AND CULTURAL ORIGINS

G. LORING BRACE, University of Michigan

PRENTICE-HALL, INC., Englewood Cliffs, New Jersey 07632

Library of Congress Cataloging-in-Publication Data

Brace, C. Loring.
The stages of human evolution.

Bibliography: p. 145
Includes index.
1. Human evolution. 2. Man, Prehistoric.
3. Anthropology, Prehistoric—History. I. Title.
GN281.B67 1988 573′.3 87-6989
ISBN 0-13-840166-7

Editorial/production supervision and
interior design: Serena Hoffman
Cover design: Diane Faccone
Manufacturing buyer: Harry P. Baisley

A Division of Simon & Schuster
Englewood Cliffs, New Jersey 07632

PRINTED IN THE UNITED STATES OF AMERICA

10 9 8 7 6 5 4 3 2 1

ISBN 0-13-840166-7 01

PRENTICE-HALL INTERNATIONAL (UK) LIMITED, *London*
PRENTICE-HALL OF AUSTRALIA PTY. LIMITED, *Sydney*
PRENTICE-HALL CANADA INC., *Toronto*
PRENTICE-HALL HISPANOAMERICANA, S.A., *Mexico*
PRENTICE-HALL OF INDIA PRIVATE LIMITED, *New Delhi*
PRENTICE-HALL OF JAPAN, INC., *Tokyo*
PRENTICE-HALL OF SOUTHEAST ASIA PTE. LTD., *Singapore*
EDITORA PRENTICE-HALL DO BRASIL, LTDA., *Rio de Janeiro*

To Mimi

whose contributions are even more than those that are so graphically apparent.

Contents

Looked at from this point of view, the Neanderthal and Pithecanthropus skulls stand like the piers of a ruined bridge which once continuously connected the kingdom of man with the rest of the animal world.

William J. Sollas, 1908

Preface

Of all the subjects that have provoked the play of human curiosity, few equal our concern with our own prehistoric origins. At the same time, few subjects have been the target for so much unprofessional speculation. Although the present work does not reduce the quantity of speculations (quite the reverse), it is my hope that they can, technically at least, bear the label of *professional* speculations. Part of the reason for the less-than-abundant work on human origins is that, in this material world, such efforts can produce little measurable gain. Physics can produce bigger and more expensive explosions, basic biology can introduce medical breakthroughs, geology and economics can contribute to our mineral and monetary well-being. But prehistoric anthropology, in contrast, can reveal only the humble nature of human beginnings, and this has dubious value as a marketable commodity. Many have regarded it as an interesting hobby, but few have been tempted to treat it as a serious career and devote lifelong concentration to its advancement. Even at the professional level, the competition to qualify has often been less than it is for other fields. Consequently, there are fewer jobs, and hence fewer practitioners, with the result that advances and accomplishments have been far less spectacular than has been the case with, for instance, genetics or electronics. The subject is fascinating nevertheless, and for the professionals, its pursuit is quite satisfying in and of itself.

It is the purpose of this volume to communicate a modicum of this interest to the reader, young or old; perhaps to kindle the spark of what might grow to be another professional career; and to add a possible modifying influence, however minor, to the understandable human tendency to magnify our present accomplishments to the point that we are inclined to forget how precarious was the very existence of our predecessors until the recent past—and perhaps may be again in the immediate future.

C. Loring Brace

chapter one

Interpreting Human Evolution

SCIENCE AND RELIGION

As a general rule, there need be no conflict between science and religion. Many scientists are deeply religious people, and, although occasionally their manifestations of religious belief are on the unorthodox side, they are frequently of one or another traditional denominational stance. The Darwinian theory of evolution also is not viewed as being incompatible with the vast majority of organized religious systems, even when evolutionary expectations are applied to the human condition. The largest Christian denomination, the Roman Catholic Church, has faced the matter squarely and has published the results of its deliberations in a series of key papal encyclicals—*Providentissimus Deus* and *Divina Afflante Spiritu,* in 1943, and *Humani Generis,* in 1950. In these, Pope Pius XII declared that the Bible was a religious and moral document and was not intended as a discourse on geology and natural history. He concluded that evolution was a perfectly valid hypothesis, and that evolutionary research and teaching were perfectly valid enterprises that did not conflict with or overlap the concerns of the church as long as they did not presume to deal with matters of morality and the human soul.

Most scientists and members of the clergy are perfectly content to leave it at that, but people in some denominations find aspects of science—particularly those that deal with human evolution—in conflict with their basic beliefs. Among these are certain "fundamentalist" Christian groups who declare that the words of the Bible recount literal truth, and anything else therefore is wrong by definition. According to them, evolution is in conflict with their religion. I can do nothing to help readers with such a starting presumption, other than to say that this book is not for them.

Of course, those who start from such a position also start with something of a conundrum, because at the very beginning of the Bible, Genesis 1 lists a very different order for creation than Genesis 2. If indeed the words of the Bible are to be taken literally, then Genesis 2 must be wrong because it contradicts Genesis 1. But if the Bible "cannot be wrong," then two successive chapters that say opposite things must both be right, even if that is logically impossible.

Although this "most ingenious paradox" may worry the literal-minded theologian, it is beyond the realm of science. So too is that contradiction in terms, "creation science"—offered by the same fundamentalist Christians as deserving equal time in the classroom with evolutionary biology. The interpretations offered by evolutionary biologists can be put to the test by experiment and the collection of relevant data, and, if they are found wanting, they are discarded. "Creation science," on the other hand, cannot be tested in such a way, and its proponents will not accept the possibility that crucial experiments or collected data could lead to its rejection. For this reason, "creation science" cannot be science.

GOD AND THE GAPS

The fossil record is admittedly incomplete, because the vast majority of the creatures that lived in the past have died without leaving a trace. The appearance of change as one ascends the picture in superimposed strata tends to be discontinuous, with gaps of varying length between the earlier and the later sections. The different appearance of sequential forms has been taken by some to be a proof for the existence of God.

This, however, has to be about as demeaning a criterion for demonstrating Divine Existence as one can imagine, and it does little credit to those who accept such a test. It purports, in fact, to use human ignorance as a proof of Divinity. When gaps are filled in, as is being done year by year, what does that do for the nature of the God who was postulated to account for them? To use gaps in our knowledge as proof for the existence of God is, in effect, to deify ignorance.

We are finite creatures, and our knowledge will always have its limits. But to deny the efforts of scientists and scholars to try to extend those limits and to glory in our imperfections is to do less than justice to our potential capacity to learn about the endless variety and mechanics of the natural world—itself something that is far more worthy of the designation "Divine."

EVOLUTIONARY EXPECTATIONS

Few among the educated and no serious scholars doubt that *Homo sapiens* evolved by natural means from a creature which today would not be considered human. From this initial point of agreement, the thinking of those who are considered qualified to judge diverges to such a degree that many feel we do not have a basis which is adequate enough to warrant any interpretation at all. Yet schemes have been constructed that attempt to arrange the prehistoric evidence and account for the course of human evolution. The pages that follow present and discuss the strengths and weaknesses of a number of these.

Since it is generally agreed that evolutionary thinking should be applied to the course of prehistoric human development, it would seem unnecessary at first glance to consider the nonevolutionary or even antievolutionary views of pre-Darwinian thought. As we shall see, however, the differences between several of the attitudes discernible at the present time can be traced in part to the lingering influence of a current of thought that has specific pre-Darwinian sources. Once this has been identified and the historical connections have been traced, the reason for the differences between the major opposing interpretations will become obvious, and we shall have some basis for making a choice between them.

Interpretations of the human fossil record can be arranged along a spectrum between two polar and opposed approaches. At one extreme is the school that takes all the known hominid fossils, arranges them in a lineal sequence, and declares that this is the course that human evolution has pursued. The other extreme is the school that declares that the great majority, if not all, of the available fossil record has nothing to do with the actual course of human evolution. The probability that any given fossil has descendants that are still alive is so vanishingly small that to declare otherwise is to be guilty of the unscientific stance of "ancestor worship." Furthermore, the course of evolution is never a straight, unbranching line—witness the diversity of related forms in the organic world today—and we should expect to find branches and specializations among human fossils. This latter view tends to regard the differing fossil hominids as "specializations" away from the main line of human evolution, which eventually became extinct without issue. And, because the chance that a given remote fossil was literally ancestral to anything still alive is clearly almost zero, it can be completely discounted for all practical intents and purposes. Countering this is the argument that, although indeed any given prehistoric individual is unlikely to have living descendants, the *population* to which it belonged certainly may be ancestral to continuing populations, and said individual is our best bet for getting a picture of what the members of that potentially ancestral population actually looked like.

People are invariably fascinated by investigating the skeletons in their closets, and, in the field of human evolution, we could say that this is *literally* the case. This fascination has led many people, amateur and professional, to write about the human fossil record—people who have not been fully qualified and who have failed to perceive the nature of the two schemes just mentioned. As a result, many authors prefer some hazy middle ground since they feel that both schemes have some merit. Consequently, few authors today represent the poles in fully developed form.

In the first edition of this book, one of the extreme positions was specifically defended—the linear scheme mentioned above—not because there was conclusive proof for it, but in an effort to follow the principles expounded by the medieval logician, William of Occam, for whom the best explanation was always the simplest. At the time, it appeared that the complexities of the most widely accepted interpretive schemes were more a product of the minds of their advocates than they were the necessary result of the available facts. Simplification, however, can be pushed too far, and the wealth of discoveries over the last twenty years have clearly shown that a rigid unilinear interpretation is, in fact, an oversimplification.

It is still true, nonetheless, that the simplest interpretation that accounts for all of the facts is the one that should be accorded top preference. And, as we shall see, the one offered here is the simplest one available. Certainly the student will discover that it is the easiest one to learn.

More of this later, but first it should be instructive to sample the various other current views on the course of human evolution. First among these, and generally regarded as most traditional, is the view that the different forms in the human fossil record are the results of the adaptive radiation of the basic human line. At present, several versions are being stressed: one that treats the entire human fossil record as a picture of divergent "specialized" lines, most of which became extinct without issue; another that concentrates on the earlier parts of the record where various "specializations" are supposed to have occurred; and finally one that concentrates on the latter part of the record where the Neanderthals are identified as "specializations" on their way to extinction. Running through all of these is the tendency to deny possible ancestral status to any fossil that differs from modern form to any marked extent. To some degree, then, these schemata focus more on assertions concerning how human evolution presumably did *not* occur than on trying to find out what was actually going on.

TIME, GEOLOGY, AND FOSSILS

To understand these applications and the criticisms that can be made of them, it is first necessary to gain some sort of perspective on the time scale in question, the fossils concerned, and the principles involved. Briefly, it has become apparent that the span during which the events of human evolution occurred was not just 300,000 or 800,000 years, as was once believed, but somewhat more than 3 million years in duration. Previous estimates were based largely upon guesswork involving sedimentation rates and stratum thicknesses. This recent reappraisal is derived from the work of geophysicists who have utilized the known and constant rates of decay of radioactive elements into their stable end products, especially the decay of Potassium 40 into Argon.

The Potassium-Argon (K/A) proportion in ancient volcanic rocks is directly related to the length of time since they have cooled, and, although many pitfalls are connected with the use of this technique to date strata in the recent past (3 million years is dewy fresh in the full perspective of geological time), it is becoming increasingly apparent that the duration of existence of the human line has been sufficient that we need not invoke an unusual rate of evolution to account for all the changes the human fossil record reveals.

The geological period during which human form did most of its evolving is called the Pleistocene, which extended from about 2 million years ago to 10,000 years ago—if indeed it can properly be considered to have ended. The oldest of our close fossil relatives are found in the Pliocene, some 3 million and more years ago, and are referred to as Australopithecines. These flourish for a span of about 2 million years, during which time they display a diversity of size, form, and robustness that has been the subject of some vigorous scholarly disagreements. The size spectrum runs from the modern average for bulk and stature down to creatures only half as large. The earliest ones appear to be small, and the most robust ones appear to be late, but it is abundantly clear that little ones continued to exist at the same time that big ones flourished. Aside from their sometimes different bulk, the most evident points of distinction between the Australopithecines and modern humans are in the head and face. Simply stated, the Australopithecine head is smaller—the brain is scarcely more than a third the size of the modern average—while the faces and teeth are enormous.

At the moment, there is a healthy professional brawl going on over the relationships between the various robust and gracile Australopithecines to each other and to the larger continuing picture of human evolution. A solution to the controversy is suggested later in this book, but for now this initial brief sketch is offered so that the reader can have some framework on which to arrange the arguments that follow.

Another cluster of hominid fossils, which can be called Pithecanthropines, dates from the middle of the Pleistocene, some half a million years ago. Among these fossils, brain size is double that of the Australopithecine and about two-thirds that of the modern norm. Molar tooth size has dropped markedly from Australopithecine levels, and a remarkably robust skeletomuscular system is maintained. In the Upper Pleistocene, immediately prior to the appearance of people of recognizably modern form, there is a fossil group which has been called the Neanderthals. These are characterized by the achievement of fully modern levels of brain size but the preservation of most of the rest of the characteristics of the Middle Pleistocene Pithecanthropines. To be sure, other fossils are unevenly scattered, in both the geographical and temporal sense, which provides a source for much of the disagreement that still surrounds any attempt to develop a systematic view of human evolution. But the foregoing should provide a useful outline to remember while the discovery of the human fossil record is being recounted.

The scheme that is developed in later chapters essentially takes these major blocks of fossil hominids, arranges them in a temporal sequence, and explores the evolutionary logic that can be used to show how the earlier ones evolved into the later ones. These major groups form the evolutionary stages through which it is claimed that the human line passed. Yet it should also be remembered that the identification of these supposed "stages" is largely dependent upon the accident of discovery. A few rich sites have provided concentrated evidence for particular forms of human fossils, and it is not only possible but extremely likely that, had these rich sites involved different time levels, then the identification and number of important stages in human evolution would have been rather different. On the other hand, the present stages perceived are adequate to represent the changes involved, and their consideration can be justified in terms of their utility.

Ultimately, when the entire time spectrum of human existence is documented by an as-yet-unforseeable abundance of fossil evidence, the picture should be one of a gradual continuum of accumulating change, with no visible breaks between what are here considered as stages. Even this, however, is vigorously disputed by one currently popular set of theoretical expectations, but this will be treated in a later chapter. In contrast, the view presented here is that human evolution has been continual in the past, it continues in the present, and it will continue in the future. Our concern in this book, however, is with the changes that have taken place in the past. One of the principal objects of this book, then, is to attempt to apply Darwinian principles to a field that has often honored his name while neglecting the use of his perspective.

The first concern, however, is to examine the sequence of discoveries and interpretations that led us to the position in which we now find ourselves. Interpretations of the major pieces of evidence are heavily conditioned by the attitudes prevalent at the time and place of their discovery. Traditions of interpretation, once established, tend to continue, whether or not subsequent evidence provides justification. The story of how we became aware of the evidence for our predecessors, who made the discoveries, and what they thought it all meant is a fascinating one in itself. And in its telling, we can come to see how the various modern interpretations arose. This, then, is the purpose of Chapters 2 through 6.

chapter two

Fact and Fancy before 1860

BEFORE THE NINETEENTH CENTURY

The earliest recognition of a fossil human was accorded a skull fragment discovered in the year 1700 at Canstatt, near Stuttgart in western Germany. At this early date, however, there was not even the remotest suspicion that modern living forms, including human ones, might have evolved by natural means from earlier forms ultimately quite different in appearance. Nor was there any faint hint of the vistas of geological antiquity that research was to reveal in the subsequent century. The Canstatt skull was accepted by some as evidence for human existence in ancient times, but its form was not different from that of modern human form, and "ancient times" were measured in terms of a total span since creation—thought to be somewhat less than 6,000 years.

As late as the middle of the seventeenth century, the vision of such antiquity was considered somewhat daring, although it had received a certain amount of religious sanction in the work of the biblical scholar James Ussher, the Anglican Archbishop of Armagh in northern Ireland. Computing from the named generations recorded in the Bible, Ussher arrived at the conclusion that creation had occurred in the year 4004 B.C. To this, the Reverend Dr. John Lightfoot, vice-chancellor of Cambridge University, added the pronouncement that ". . . heaven and earth, centre and circumference, were created all together in the same instant, and clouds full of water. This work took place and man was created by the Trinity on October 23, 4004 B.C. at nine o'clock in the morning."

By the end of the eighteenth century, 150 years later, appraisals of geological processes and accumulating knowledge of the structure and strata of the earth led to the suspicion on the part of some people that the earth was really very much older. Fossil remains of extinct and different animals had been discovered, and scholars were becoming aware that the world had been a very different place in ages gone by, and that great changes had occurred. A few people even noted that the shaped pieces of flint discovered in prehistoric strata might be human tools made before the discovery of metallurgy, and certainly historians and students of human institutions were aware that the human world had changed even in the recent past.

Early in the nineteenth century, the French biologist Lamarck tried to promote a view that continuous and accumulating change was the normal state of affairs. He really was a thoroughgoing evolutionist, but the mechanism that he proposed to account for organic change was invested with an element of unscientific mysticism, and his position has subsequently been generally rejected. The initial reason for this rejection was the fact that many people were emotionally unprepared to accept change as the normal expectation. The traditional view that the world was created fixed and changeless had both social and religious support, and a scheme proposing the normality of constant change was regarded as a threat to the established order. Yet change could be seen in the geological record of the remote past, and some sort of explanation was demanded.

CUVIER AND CATASTROPHISM

An acceptable solution was proposed by another French scholar, Georges Cuvier, who was a younger contemporary of Lamarck. The English philosopher of science William Whewell coined the term "catastrophism" to describe Cuvier's scheme. It claimed that the various geological layers had been deposited as the results of a series of cataclysms that had overwhelmed the planet periodically, extinguishing all previously living organisms. According to this scheme, the last of these cataclysms was the Biblical flood, which meant that human remains should not be discoverable in earlier layers. Cuvier is credited with the statement: "Fossil man does not exist." And indeed, in the early nineteenth century there was very little known evidence to contradict such a position.

Cuvier was somewhat vague concerning the origin of the new animals that appeared in the strata overlying his various supposed cataclysms. Such cataclysms need not have been worldwide, he noted, because the detailed geological sequence in one part of the world tended to differ from the specifics of sequences in other parts of the world. Following the presumed cataclysm that eliminated the living forms in one given area, then, he suggested that those which had continued to live in places not so affected might just migrate into that now-empty region. Ultimately, however, in order to account for the disappearance of all former types of organisms and the continued emergence of the new and different, Cuvier's stance gave support to a philosophy of successive creations. With the development of Darwinian evolutionary theory in the middle of the nineteenth century, the view of supernaturally caused extinctions, migrations, invasions, and successive creations was superseded as a general explanation. Yet, because of a variety of historical accidents, something of this has survived into the present in the traditions of paleontology—especially that part of paleontology particularly concerned with the matter of human origins: paleoanthropology.

Georges Cuvier (1769–1832), zoologist, comparative anatomist, paleontologist, and unwitting influence on many of the subsequent attempts to interpret the human fossil record. (Brown Brothers.)

The discovery of the fossil and archaeological evidence for human evolution was the result of the fieldwork of people who had very little concern for the research that developed the evolutionary explanation for the origin of organic diversity and organic change, yet both realms of activity have parallel careers extending back into the eighteenth century. Archaeological and paleontological work could and did go on without much concern for theoretical implications. Even though Cuvier was specifically opposed to evolution, he can be regarded as the founder of paleontology, a discipline that, ironically, provides the most direct evidence in support of evolutionary theory. His intellectual descendants (and other unrelated antiquarians and archaeologists) pursued their diggings right up into the twentieth century, often with quite incorrect assumptions concerning their interpretations; today these figures are the principal sources of the other interpretations of the human fossil record that we discuss in later chapters. Darwin, on the other hand, used relatively little paleontological evidence to support his major insights. This was partly because of the very incomplete nature of knowledge concerning the fossil record, and partly because his concern was focused on the attempt to explain diversity in the world of *living* organisms.

DARWIN AND THE ORIGIN

Although it has remained for the twentieth century to attempt a synthesis of the study of the present with the study of the past, scholars in both areas have not been unaware of the implications each has had for the other, and the public has been sensitive to this from the beginning. This still shows in the common misconception concerning the title of Darwin's most famous book, *On the Origin of Species.* From the time of its appearance right up to the present, people who were not thoroughly familiar with the book assumed that it suggests a common ancestry for apes and people, and that the "species" in the title refers to humanity itself. This latter assumption is so strong that the title is frequently misquoted, as *The Origin of* **The** *Species.* Actually, only one brief sentence at the very end of the book makes any reference to humans at all, and this is thoroughly noncommittal. Darwin's concern for human evolution was reserved for expression in another book, *The Descent of Man,* published more than a decade after his *Origin.* Even here, however, his reference to the skimpy fossil and archaeological record of human prehistory is brief in the extreme.

Charles R. Darwin (1809–1882), author of *On the Origin of Species* and acknowledged father of evolutionary thinking. (National Portrait Gallery, London.)

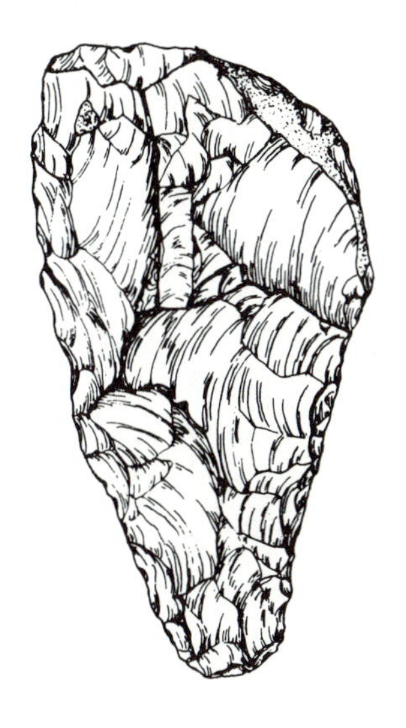

A biface from St. Acheul in northwestern France, the location that gave its name to a whole category of middle Pleistocene tools. (Drawn by M. L. Brace.)

ARCHAEOLOGY

The trickle of accumulating evidence had been growing, however, with prehistoric skeletons and stone tools being brought forth even during Cuvier's lifetime. During the 1820s, human skeletal material was discovered in association with extinct animals and ancient stone tools on the coast of Wales, in France, and in Belgium, but none of it attracted much attention. Late in the 1840s, Boucher de Perthes, a customs inspector at Abbeville in northwestern France, published the results of his prehistoric investigations, over the previous fifteen years, at both Abbeville and St. Acheul. In the gravel of the Somme River terraces Boucher de Perthes had discovered flints of such a regular shape that they could only be the products of human manufacture. Yet they obviously were deposited during the course of the formation of the terraces in which they were found, which suggested an age for their makers far in excess of anything granted by even the most liberal supporters of human antiquity. Similar stone tools have since been discovered in prehistoric strata elsewhere in Europe, Africa, India, and Asia. Today the name Acheulean is used to refer to the type of artifact first recognized by Boucher de Perthes at St. Acheul.

GIBRALTAR

Relegation to unimportant obscurity was very nearly the fate of the archaeological discoveries of Boucher de Perthes. Contemporary French scholars were so scornful of his claims that they never even bothered to visit his diggings or investigate his work firsthand, but simply remained in Paris and denounced him from a distance. Had it not been for the curiosity of a group of English scientists who visited his sites and confirmed his findings, these artifacts would have had as little influence on the study of human origins as had the Gibraltar skull, which was found just a year

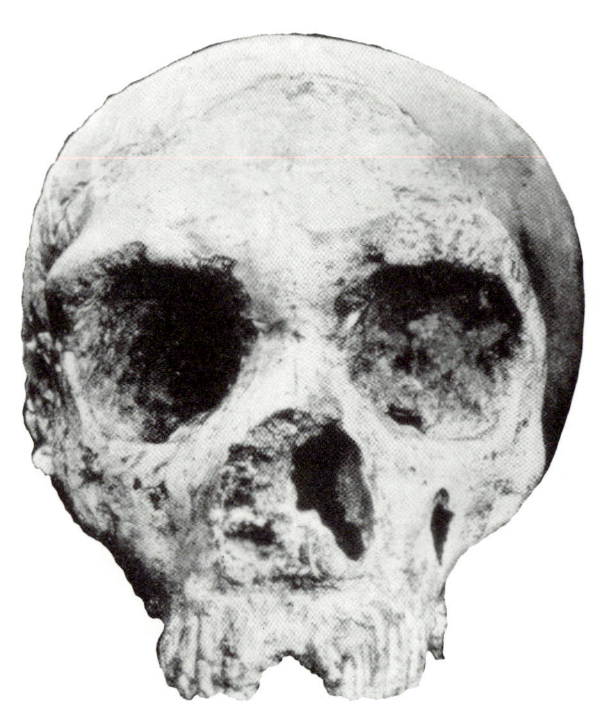

The Gibraltar discovery—a skull of Neanderthal form.

later, in 1848, in a quarry on the north face of the Rock of Gibraltar. The discovery of this skull, which we now recognize as a representative of the Neanderthal stage of human evolution, was recorded by the Gibraltar Scientific Society. A slow fourteen years later, after the skull had found its way to England, it was shown to at best mildly interested scholars at meetings of the British Association for the Advancement of Science and at an anthropological congress. Just twenty years after its discovery it was presented, pretty much as a curiosity piece, to the Museum of the Royal College of Surgeons in London, where it remained, almost forgotten, until after the turn of the century. Because its importance went unappreciated for more than half a century following its exhumation, it played no part in the development of the study of human evolution, which is rather a pity since it differs markedly from the stereotype of the heavily buttressed, muscle-marked, and robust image that is conjured up in so many minds when the name Neanderthal is invoked. The Gibraltar skull, as it happens, is that of a female in a stage in which male-female differences were more marked than is now usually true.

NEANDERTHAL

By a remarkable set of historical coincidences, the late 1850s saw the discovery of the skeletal remains of what could be identified as an earlier stage in human evolution, the recognition of the archaeological evidence for human antiquity, and the development of an intellectual framework within which these new facts could be encompassed. The specific timing of these events was a little less fortunate, because the discovery of the skeletal remains occurred first, and they were the subject of skeptical comments that have influenced interpretations ever since. The skeleton was discoverd in 1856 during quarry operations in a limestone gorge through which flows the Düssel River, a tributary of the lower Rhine. The gorge lies in the area between Elberfeld and Düsseldorf and bears the name of Neanderthal. By giving its name to the skeleton discovered there, it has provided a designation for the entire stage of evolutionary development immediately prior to the emergence of modern human form.

The skeleton had evidently been a burial in a small cave in the limestone cliffs, and had probably been complete. In the course of being recovered it was somewhat battered, because it, along with the dirt in which it lay, was unceremoniously shoveled out onto the terrace by quarry workers who were cleaning out the cave to get at the rock. Its human nature was only recognized later by Johann Karl Fuhlrott, a natural science teacher at the high school in Elberfeld, who assured its preservation. Fuhlrott, with the aid of Hermann Schaaffhausen, a professor of anatomy at Bonn, promoted the view that this was an early form of human, but this interpretation received no support until very nearly the end of the nineteenth century. Possibly because of the mode of excavation, the face was not recovered. The head was represented by the skullcap from the ridges over the eye sockets extending to the back of the skull but minus the base. The limb bones were extraordinarily robust and the brow ridges of the skull enormous, but, lacking the face, jaws, and teeth, the evidence for clear difference from modern form was subject to debate.

And debate there was. Enough peculiarities were present to suggest all sorts of explanations, from derogatory hints that the specimen was an ancient Celt of "low type" similar to the modern Irish, to suggestions that it was an idiot, a freak, the victim of rickets, or the residue of the Mongolian Cossacks who had chased Napoleon back from Russia in 1814. The most authoritative opinion was delivered by one of Germany's leading scientists, Rudolf Virchow, a founder of German anthropology and, as the originator of the field of cellular pathology, the most outstanding pathologist of the day. After careful examination, Virchow pronounced the skull "pathological" and sought to explain all of its peculiarities in that fashion. The weight of his judgment has been such that Neanderthal morphology has been regarded as "aberrant" from that day to this, and a majority of authorities even today refuse to accept Neanderthals as representative of the ancestors of modern human beings. In contrast to all of this, the argument is developed in a later chapter of this book that Neanderthal form is just what we would expect to find prior to modern emergence, and Neanderthals are the most logical representatives of modern ancestors.

In 1858, the year following the first discussion of the Neanderthal discovery, a delegation from the Royal Society of Great Britain visited the excavations of Boucher de Perthes in northern France and returned, convinced of the significance of his work, to report to the British scientific world. Then, in 1859, Darwin's book *On the Origin of Species* appeared. From then on, human attitudes toward the world of nature and their own position within it were permanently altered. No longer could people regard themselves as the epitome of existence in a world created solely for their own benefit. Of course, for quite some time, in the pride of their self-importance, many people could not accept the implications of this presentation. However, as time went on and acceptance became nearly universal, it became apparent that the consequent enforced humility was doing people no harm. The upshot of the entire matter is that no area of human behavior and philosophy has escaped the impact of the consequent revolution in attitudes.

chapter three

The Picture up to 1906

CRO-MAGNON

The vindication of Boucher de Perthes and the intellectual revolution going on in Britain could not fail to be a great stimulus to prehistoric research. During the late 1850s and 1860s, basic work on discovering the characteristics of human cultures prior to the existence of metal was undertaken. In France, particularly in the Dordogne region and the Vézère River valley of the southwest, excavations at La Madeleine, Solutré, Aurignac, and Le Moustier uncovered the evidence for prehistoric stone tool-making traditions. These were named Magdalenian, Solutrean, Aurignacian, and Mousterian and are now known to be roughly 15, 20, 30, and 40 and more thousand years old, respectively. It was suspected that they dated from a period more recent than those discovered by Boucher de Perthes, but no one then imagined that the difference was actually more than 100,000 years. Then, most exciting of all, in 1868, human skeletal remains were discovered in the same stratigraphic level with tools of Aurignacian type. To the interest as well as the relief of the public, these remains indicated that the individuals in question were not markedly different from modern human form. In fact, their appearance has been portrayed in an idealistic manner and with a glowing enthusiasm not entirely warranted by their somewhat fragmentary condition.

The human skeletal remains that we have been discussing, representing some five individuals, were discovered during the course of constructing the railroad through Les Eyzies, in the aforementioned Dordogne region of southwestern France. The removal of fill for the abutments of the railway bridge revealed a long-hidden rock shelter near an eminence called Cro-Magnon, within which the skeletons and artifacts were found. Competent geologists were on hand to verify the antiquity and the stratigraphic associations, and the study of human fossils was finally given its first solidly documented specimens. Stature of the male

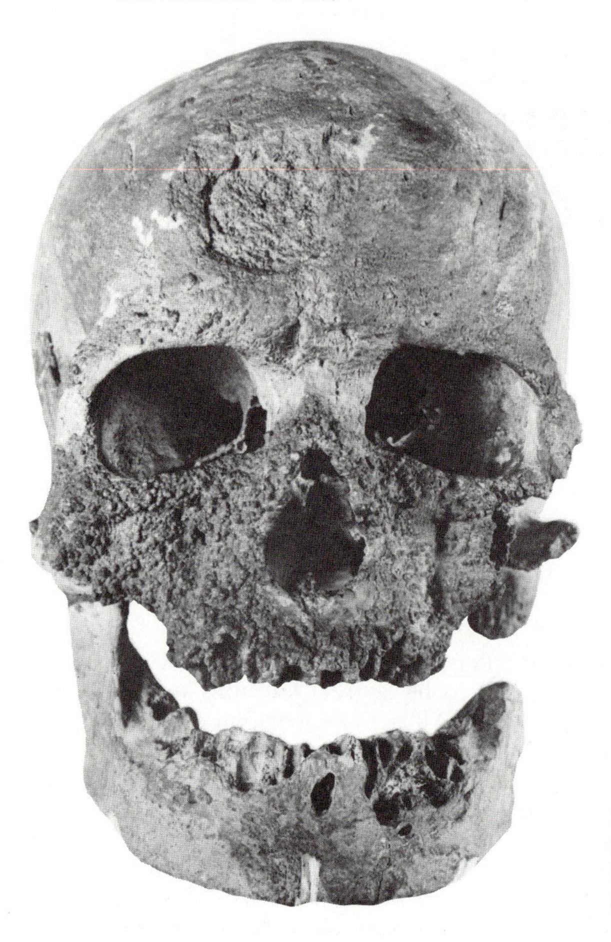

The male Cro-Magnon, from Les Eyzies, Dordogne, France. (Courtesy of the Musée de l'Homme, Paris.)

skeleton can be calculated to be 5 feet 10 inches or 5 feet 11 inches, which is tall in comparison to present or previous worldwide general averages, and the rugged long bones suggested a robust and muscular people. The face was vertical rather than projecting, and possessed a prominent chin, although these features have been stressed to a greater extent than the evidence warrants, considering that the large male cranium was toothless at discovery.

Still, the Cro-Magnon finds were recognizably of modern form and provide the basis for the still valid assertion that the Upper Paleolithic tool-making traditions—Aurignacian, Solutrean, Magdalenian, and others—were the products of people not unlike ourselves. Associated artwork in the form of carvings and engravings on bone and ivory revealed a degree of sophistication in these Upper Pleistocene hunters that was quite gratifying from the point of view of the people who were beginning to accept these Upper Paleolithic people as ancestral to more recent humanity. The decades that followed witnessed the discovery of abundant additional support for the picture outlined in the discoveries at Cro-Magnon, and it was some time before the unearthing of more ancient remains again forced people to face the issues of human evolution: the possibility that humans had arisen from something quite different from their present form.

SPY

Exactly thirty years after the original discovery in the Neanderthal, and long enough later for the controversy to have died down, two human skeletons were found buried in a Mousterian level in a cave in the commune of Spy (pronounced Spee) in the province of Namur, Belgium. The form of both skeletons was recognizably similar to that of the original Neanderthal; the skull Spy I, in fact, was of practically identical shape. No longer was it possible to expound with such certainty the supposed pathological features of the individual from the Neanderthal. However, the adamant Virchow refused to back down and, although the Spy discoveries confirmed the Neanderthaler as a valid human type, the implications of abnormality and peculiarity tended to remain. Indeed, to this day they have not been fully shaken off. Nevertheless, the Neanderthals could now be regarded as a "type" associated with a definite tool-making tradition, and they could thus be given a definite age.

EUGENE DUBOIS AND *"PITHECANTHROPUS"*

The scene was shortly to shift to another part of the world and involve another, possibly even more dramatic, form of human fossil. The German naturalist Ernst Haeckel, greatly excited by the implications of Charles Darwin's work, was communicating his enthusiasm for the evolutionary viewpoint to a rising generation of students on the German academic scene. Pushing evolutionary logic to its conclusion, Haeckel drew a hypothetical family tree linking modern humans to a common ancestry with the living apes and monkeys. He further suggested that somewhere in between the two, back in the remote past, there must have been a form which was neither one nor the other—a completely transitional stage. This, he suggested, should be referred to as *"Pithecanthropus alalus"*—that is, ape-man without speech. An American journalist was later to christen this the "missing link," a term which remains a firm item of popular folklore.

Whereas Darwin had suggested that Africa was the most likely place to search for the earliest human ancestor, because he believed that the gorilla and the chimpanzee were our closest living relatives, Haeckel and others in Germany stressed Southeast Asia, since they claimed that the detailed morphology of the gibbon's skull was more akin to the human than was that of the African anthropoid apes. Today it appears that Darwin's suspicions were sounder, although the discoveries of the 1890s made Haeckel's guess seem little short of inspired.

Fascinated by Haeckel's hypothetical portrayal of human ancestry, a young Dutch doctor, Eugene Dubois, went forth from Europe to the home of the gibbon in Southeast Asia with the avowed intention of finding the "missing link." At the time, this seemed like the most hare-brained thing in the world to do. Not only did Dubois have to give up a promising career as a teacher of anatomy at Amsterdam, but also there was virtually no shred of evidence in support of his scheme. Nevertheless, fortune smiled on Dubois, and, by a piece of impossibly good luck, he did indeed find what he was looking for. The improbability of his venture can be compared to the only other instance in which theoretical expectations launched an expedition to demonstrate the locale of human origins, which one eminent scholar felt should be somewhere in eastern Asia, the result of which was the discovery of dinosaur eggs in the Mongolian desert!

Plagued by a lack of funds—his project sounded so absurd that no one was willing to back him—Dubois signed on as a health officer in the Dutch colonial forces in Indonesia, which at that time was referred to as the Dutch East Indies. He was first assigned to Sumatra, where he spent several years hunting fossils. A variety of circumstances led him to suspect that Java was a more likely area, and in 1889 he got himself transferred there. He remained in Java for the next five years, and there he made the discoveries for which he will always be remembered. In 1890 he discovered a small fragment of a lower jaw whose importance was only recognized later. In 1891 his excavations unearthed a skullcap with such a low forehead and heavy brow ridge, and with such marked constriction between the brow and the brain case, that he attributed it to a chimpanzee. In 1892, some 50 feet away, he found a femur (thighbone), which was practically indistinguishable from the femur of a modern human. This, he claimed, belonged to the individual represented by the skull, and for a while he believed that he had discovered an ancient erect-walking chimpanzee. His comparative studies and measurements forced him to alter his opinion, because the skull, however primitive or ape-like it appeared to be, was halfway between that of a human and that of a chimpanzee in gross size. It possessed a brain that fell within the lower limits of the normal modern range of variation. He realized this was indeed his "missing link," but, in contrast to the semierect posture which had been attributed to the Spy and Neanderthal finds, it was an erect-walking "missing link." So Dubois slightly modified Haeckel's designation and, in his monograph of 1894, christened his discovery *"Pithecanthropus erectus."* This still serves as the type specimen for our Pithecanthropine stage, although it is no longer regarded as a valid separate genus.

Dubois's admirable monograph created an international sensation, and when he returned to Europe in 1895, he was an immediate celebrity. At the International Zoological Congress meeting at Leyden in 1895, Dubois and his Pithecanthropus were the focus of attention of an unparalleled gathering of famous scholars. After prolonged argument, three schools of thought emerged. One, siding with Dubois, felt that Pithecanthropus was neither an ape nor a human but a genuine transitional form. Another group felt that it was on the human side of the boundary—primitive, perhaps, but hominid nevertheless. The third group, headed by the aged Virchow, regarded it as being a giant form of gibbon, interesting and unusual, but only an ape after all.

The controversy continued unresolved for many years, and it was not until the late 1920s and 1930s, when more Pithecanthropine skeletal remains were discovered in China near Beijing, and also in Java, that general acceptance was possible. The Chinese Pithecanthropines, originally christened *"Sinanthropus pekinensis"* in 1927, were associated with stone tools and charcoal remnants once thought to be the remains of hearths, and this was accepted as evidence confirming the human status of the Pithecanthropines as a whole. Paradoxically, among the very few voices now raised in opposition to the human status of *"Pithecanthropus"* was none other than that of the elderly Dubois himself. Although he was willing to accept the newer discoveries in Java and China as genuine early human beings, he reverted to the opinion expressed by Virchow in 1895 that his own original discovery had been just a giant gibbon.

GUSTAV SCHWALBE

Although full confirmation for the significance of *"Pithecanthropus"* had to wait some thirty years after the original announcement, most scholars at and following the turn of the century came to feel that it could be regarded as an extremely primitive form of humanity. What with Pithecanthropines, Neanderthals, and Moderns established at different times, and at least the latter two associated with different archaeological traditions,

Gustav Schwalbe (1844–1916), Strassburg anatomist and physical anthropologist, who first arranged the known human fossils in an evolutionary sequence. (Courtesy Ashley Montagu.)

it was possible during the first years of the twentieth century to suggest a logical evolutionary scheme containing all the known human fossils arranged in terms of relationships and chronology. This was done by Gustav Schwalbe, professor of anatomy at the University of Strassburg, who capped a series of papers and monographs in the late nineteenth and early twentieth centuries with his summary work, *Studies on the Prehistory of Man,* published in 1906. In this Schwalbe tentatively proposed a picture of human evolutionary history comprising three successive stages—Pithecanthropus, Neanderthal, and Modern; he also allowed for the possibility of adjustments and modifications that future finds would make inevitable.

Schwalbe's scheme was useful, flexible, and in accord with the evidence available at that time. With one major addition, it has proved to be valuable enough to provide the organizing principle behind the interpretations offered in the later chapters of this book. For reasons we consider in the pages that follow, it has, however, been generally rejected and forgotten by the anthropological world.

chapter four

Hominid Catastrophism

HEIDELBERG

In 1907, the year following Schwalbe's summary, a brief wave of excitement surrounded the discovery of an enormous hominid mandible in a gravel pit near the village of Mauer, not far from the city of Heidelberg in western Germany. Without the rest of the skull, interpretations were somewhat inhibited, although the primitive characteristics were obvious. Still, the stratigraphy was precisely documented, and it indicated that the Heidelberg jaw was, as it remains today, among the oldest of the human fossils discovered in western Europe. It was a probable representative of the Pithecanthropines. Certainly it was a contemporary of those Far Eastern specimens that have given us our most detailed knowledge of the appearance of the Pithecanthropine stage of human evolution.

LE MOUSTIER

In 1908, however, the scene of discovery shifted, and a tide of historical accident began that is largely responsible for the present interpretations of human evolution in general and of the Neanderthals in particular. At

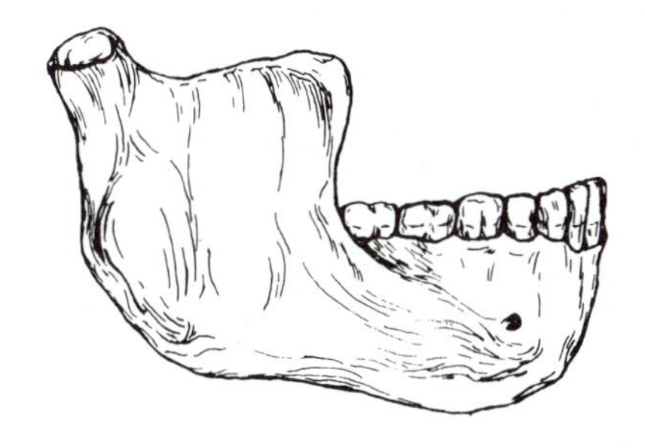

The Heidelberg jaw. Until the rash of discoveries of the last few years, this had been the only Pithecanthropine known in Europe. (Drawn from a cast by M. L. Brace.)

Le Moustier in southwestern France—the same village that gave its name to the tool-making tradition associated with the Neanderthals—a genuine Neanderthal burial was discovered. For a variety of reasons (initially related to the somewhat dubious activities of the discoverer, a Swiss dealer in antiquities who had been looting French archaeological sites and selling the booty to the highest bidder), the description was delayed for many years, and, as a result, the Le Moustier skeleton never played the role it deserved.

LA CHAPELLE-AUX-SAINTS, MARCELLIN BOULE, AND THE NEANDERTHAL CARICATURE

Later in the same year and not far from the same region in southwestern France, another and more complete Neanderthal skeleton was discovered in excavations near the village of La Chapelle-aux-Saints. These remains were entrusted to Marcellin Boule, the renowned paleontologist at the National Museum of Natural History in Paris. During the next five years Boule produced a series of scholarly papers, climaxed by a massive monograph in three installments, appearing in 1911, 1912, and 1913. Boule's portrayal of this, the most complete Neanderthal skeleton yet to have been discovered, formed the basis for the caricature of the "cave man" espoused by an entire subsequent generation of cartoonists, journalists, and, alas, professional scholars. The "Old Man" of La Chapelle-aux-Saints was depicted as a creature structurally intermediate between modern man and the anthropoid apes.

The great toe was presumed to diverge, hinting that it still preserved a degree of opposability to the other toes, and, in doing so, it presumably forced the possessor to walk on the outer margins of the feet in the awkward manner of the modern orang. Details of the knee joint were taken to indicate that it could not be entirely extended, meaning that the Neanderthals were not completely erect and could do no better than to shuffle along with a bent-knee gait. This was also supposed to indicate their similarity to modern apes, although since apes are perfectly capable of fully extending their legs, such claims demonstrate an ignorance of the anatomy and functioning of the knee joint in apes as well as humans. The same issues had been raised concerning the Spy skeletons, and several detailed studies before the end of the nineteenth century demonstrated how inapplicable they were, but Boule chose to ignore these. In harmony with the semierect picture conjured up by his discussion of the feet and legs, Boule claimed that the reverse curves present in the human neck and lower back were absent, as in the modern apes, and that the whole trunk indicated a powerful but incompletely upright postural adaptation. On top of this scarcely human caricature was a head which hung forward instead of being balanced on top of the spinal column. A detailed study was made of the cast of the interior of the braincase, and this convinced Boule that the brain was inferior in organization to that of the modern human condition, particularly in the frontal lobes, which, since the days when phrenology had been respectable, everyone "knew" to be related to the higher functionings of the mind.

The significance of continued and repeated use of such words as

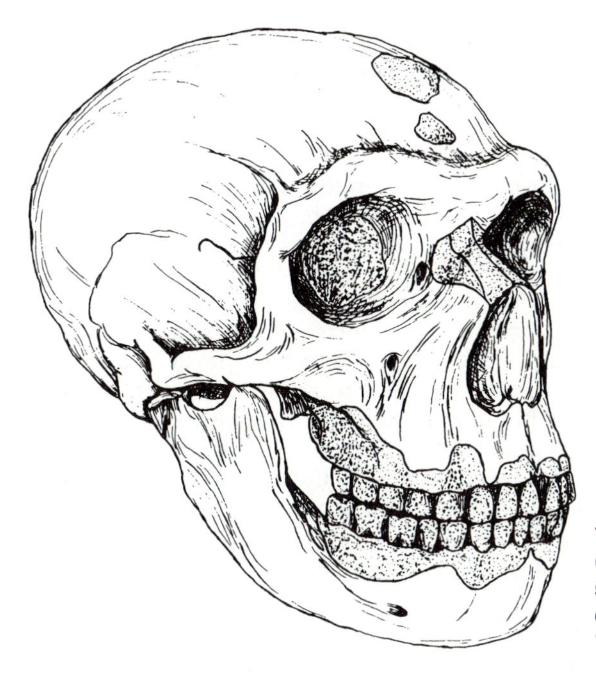

The "Old Man" from La Chapelle-aux-Saints, Corrèze, southwestern France. An extreme example of the "classic" Neanderthals.

"ape-like," "primitive," and "inferior" was not lost on the fascinated public, which quickly invested the Neanderthals with a veritable hairy pelt and long simian arms, although there is no evidence whatever concerning hair, and the arms were actually relatively short. In the years since that time, it has been demonstrated that Boule was in error on each one of the foregoing points, but the vision of the totality has not been altered, and the Neanderthals continue to shuffle through the pages of numberless books and to slouch stupidly in countless cartoons.

Having produced this caricature, Boule then proclaimed that it could have nothing to do with the ancestors of modern *Homo sapiens.* As justification, he claimed that the Neanderthals and their culture came to an abrupt end and were suddenly replaced by full *Homo sapiens,* who swept into Europe with their superior Upper Paleolithic technology. Furthermore, said he, people of modern form already existed when the Neanderthals were the main occupants of the European scene. This latter claim has provided one of the main stimuli for subsequent activities in human paleontology because, within a year, Boule's candidate for a representative proving the existence for this ancient Modern was disqualified. From that time on, two subsequent generations of anthropologists have been searching for the as-yet-undiscovered *sapiens* of modern form in the Middle Pleistocene or even earlier.

THE PILTDOWN FRAUD

Within the same year that the final installment of Boule's ponderous work appeared, an ingenious Englishman fabricated the famous Piltdown fraud, which confused the picture for a full forty years before being exposed.

Piltdown turned out to be the fragments of a modern human cranium and part of the jaw of a modern female orang, stained to look ancient, appropriately broken and artificially worn, and mingled with a collection of extinct animal bones acquired from all over the world before being scattered in a gravel pocket in southeastern England.

Also in the same year, Gustav Schwalbe published a review of Boule's monograph that, at eighty pages, was nearly a book in itself. In this he yielded to the picture painted by Boule and abandoned his own former claims that the Neanderthals were the direct ancestors of modern human beings, even though he noted the errors in Boule's treatment of Neanderthal anatomy and the fact that the evidence did not support Boule's claim for the existence of ancient Moderns. Schwalbe never abandoned Dubois's Java discovery, which he continued to regard as ancestral to all later human forms, although Boule had indicated that he considered *both* the Pithecanthropine and Neanderthal groups to be branches off to the side of the mainstream of human evolution—branches that became extinct without descendants. Curiously, despite the continuing lack of evidence to support it, the interpretation proposed by Boule remains the majority view right up to the present day among professional paleoanthropologists.

THE IMPACT OF WORLD WAR I

In 1914, the year after Boule's final publication on La Chapelle-aux-Saints, World War I burst upon Europe. Dislocation of human affairs and cessation of scholarly activity are inevitable companions of war, but in the field that pursues the study of human evolution, the legacy of this conflict has been more enduring, if less clearly appreciated. Germany not only lost the war, but suffered a blow to its intellectual prestige that has had repercussions ever since. In the postwar era, Germany's intellectual recovery was progressively stifled by the rise of the Hitler regime, which, when it came to power, quickly extinguished what had managed to survive. This was particularly true of any science that attempted to make an objective and unbiased study of human beings. Anthropology and the other social sciences suffered severely. They have made few contributions, and they have played but a minor role in the subsequent general advances made in other countries. Little remains of the pre-World War I tradition in German anthropology and, while I can hear my colleagues muttering that this is really a good thing, the valuable parts have been eliminated along with the bad ones. It is to be regretted, for instance, that so little is remembered of the pre-Boule writings of Gustav Schwalbe.

HOMINID CATASTROPHISM

Before proceeding, it is of more than idle concern to note the source of Boule's orientation. Boule was a paleontologist, trained during the 1880s in an academic environment that had not accepted the Darwinian view of evolution. Although French paleontologists spoke of "evolution," they carefully distinguished it from "Darwinism." To them, evolution signified the appearance of successive organic forms, whereas "Darwinism" meant the development of later forms out of earlier ones by natural processes, and this they refused to accept. When questioned concerning the source

Aleš Hrdlička (1869–1943). Born in Czechoslovakia, raised in the United States, he was the first physical anthropologist at the Smithsonian Institution and one of the most distinguished representatives of the field in America. Hrdlička was one of the very few scholars after the First World War who continued to view the Neanderthals as a stage in human evolution. See his *The Skeletal Remains of Early Man,* 1930. (Courtesy of the Smithsonian Institution.)

of the successive forms, they would evoke extinctions followed by invasions from elsewhere, and, ultimately, successive creations. This, then, was simply the survival of Cuvier's "catastrophism," relabeled "evolution," and it was what Marcellin Boule applied to the human fossil record. As he noted, modern-looking humans appeared more recently than Neanderthals, so, following the tradition in which he was trained, he postulated Neanderthal extinction and subsequent modern invasion. This, of course, presupposed the existence of modern forms elsewhere, about which he, like Cuvier a century before, was relatively vague. Such a presupposition has caused his followers a considerable degree of mental anguish ever since.

This view can be labeled hominid "catastrophism," and, because of the historical accidents of the second decade of the present century, it has continued to dominate interpretations of human evolution, although the recognition of its intellectual roots and original justification has been largely forgotten. Following World War I, Boule treated the totality of the known human fossil record, presenting his scheme of hominid catastrophism—labeled "evolution"—in a single volume, *Fossil Men.* This book continued to influence the field long after Boule's death in 1942. It was revised by his student and follower, H. V. Vallois, after World War II and enjoyed considerable popularity in its English translation. And recently, the intellectual stance of pre-Darwinian paleontology and biology has gained a wide following in the guises of "cladistics" and "punctuated equilibria." We treat these matters later, in the chapter on evolutionary principles.

chapter five

Between World Wars

RAYMOND DART AND *AUSTRALOPITHECUS*

Over the years, far more fossils have been discovered than there is room even to begin to record in a treatment as brief as this. Some of these, relatively important specimens, must be omitted because they have not met the criterion of contributing substantially to alterations in the overall picture. The next discovery that did measure up was made in 1924. Although it was but a little fossil for which only modest claims were made, in retrospect it can be regarded as a major portent of what was to come.

The scene was South Africa, where a small fossil skull was given to the young Professor Raymond Dart at the medical school of the University of the Witwatersrand in Johannesburg. Dart had recently finished his training in medicine, anatomy, and physical anthropology in London, and he was keenly aware that Africa, cited by Darwin as the possible source of the human line of development, had up to that time yielded no dated early human material at all. Only the single enigmatic and undated find of a Pithecanthropus-like skull in a mine shaft at Kabwe (then Broken Hill) in Zambia (formerly Northern Rhodesia) existed to demonstrate the presence of an earlier stage in human evolution.

The little skull handed to Dart was that of an immature individual—approximately at the same stage of development as that of a modern 6-year-old child—and it is risky to establish taxonomic affinity or evolutionary stage on the basis of specimens in which growth has not been completed. Still, Dart's study, published early in 1925, was able to demonstrate that his juvenile creature had a brain the size of an adult gorilla's, that its head was balanced atop the spinal column instead of slung forward, that the palate was human rather than ape-like in shape, and that, despite the great size of the teeth, the canines did not project beyond the level of the other teeth. Although Dart correctly noted that juvenile apes are less distinct from juvenile humans in more of these features than the adults of the various forms, he could state in summary that this South African fossil, blasted out of a quarry at Taung, presented a curious mixture of ape-like and human features. Withal, he regarded it as an extinct ape—closer perhaps to the human line than any yet discovered, but an ape nevertheless—and christened it *Australopithecus africanus* (southern ape of Africa).

Dart's sober and relatively cautious appraisal was greeted by an outburst of patronizing scorn from the evolutionary and anatomical authorities back in England, several of whom were his former teachers. Chief among these was Sir Arthur Keith, champion of the Piltdown fraud, who repeatedly stated that Dart's position was "preposterous." In fairness, it should be pointed out that Keith was in no way to be blamed for the fraudulent facets of the Piltdown melange, because he was as badly—even tragically—misled by it as anyone. Yet of all the criticism offered of Dart's views, only a relatively trivial one remains: that he mixed Latin and Greek roots in a single term and used a substantive in place of an adjectival form in assigning the fossil its name.

There seem to have been two sources for the reaction to Dart's claims. One was the feeling that the fossil should really have been turned over the the "proper authorities" (namely, those back in England) for study. The second was based on the feeling that, with Dubois's Pithecanthropus finally accepted as the earliest possible form of human (by all except Dubois himself), anything demonstrably more primitive, as *Aus-*

Raymond A. Dart, whose prophetic interpretation of *Australopithecus* went unappreciated for more than thirty years. (Photo courtesy of Professor Raymond Dart.)

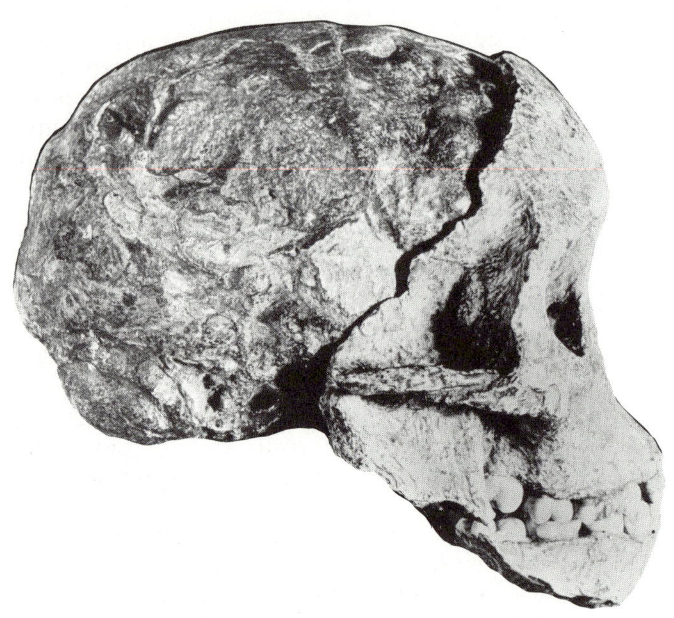

The first Australopithecine to be discovered, Dart's original *Australopithecus africanus*. (Courtesy of the Department Library Services, American Museum of Natural History.)

tralopithecus was, even had it grown to adulthood, could not conceivably belong in the picture. Furthermore, that Dart, like Dubois, should make such an epochal discovery within just a few years of his arrival seemed just too much of a coincidence to be believable. Influenced by such considerations, "Dart's child," as *Australopithecus* was deprecatingly designated, was relegated to the category of "just another fossil ape," although that alone should have been reason enough to attract attention because at that time fossil apes were—as they are even now—rare specimens indeed.

FRANZ WEIDENREICH AND PITHECANTHROPINES IN CHINA

The excitement over *Australopithecus* was soon superseded by the discovery of Pithecanthropines in China, the so-called *"Sinanthropus"* remains or "Pekin Man." During the succeeding decade, fragments of more than forty individuals were retrieved from the limestone caves of Zhoukoudian (Choukoutien), a few dozen miles southwest of the Chinese capital, Beijing (then called Pekin). Ultimately, these were the subject of a series of masterly monographs by Franz Weidenreich, a refugee from Hitler's Germany teaching anatomy at the Peking Union Medical School and, by great good fortune, one of the very few who perpetuated the outlook of his late teacher and colleague, Gustav Schwalbe.

War, with its inevitable disrupting influence, spelled final oblivion for the original Sinanthropus material. These were last seen on December 7, 1941, the day when the bombs fell on Pearl Harbor signalling the entry of the United States into the war that Japan had begun against China four years earlier. Casts of the hominid fossils had been sent to the United States previously, and the originals were in the process of being sent for safekeeping, but Japanese troops captured the American marine detachment escorting the material at the port of embarkation, Jin Wang Dao, and no trace of them has since been found. However, during the brief time of their resurrection, excellent casts, drawings, and photographs had been made—all of which survive—and the exhaustive descriptions and comparisons in Weidenreich's splendid publications have removed all remaining doubts concerning the human status of the Pithecanthropines.

The 1930s also saw important discoveries in other parts of the world. A skull from Steinheim in Germany (1933) and the back of a skull from Swanscombe in England (1935 and 1936) partially served to portray the nature of the western successors of the Zhoukoudian group, although interpretations are still the subject of prolonged debate that owes more to the continuing reluctance to accept the implications of a Darwinian viewpoint than it does to the nature of the fossils themselves.

Franz Weidenreich (1873–1948) shown with Dr. G. H. R. von Koenigswald. (Courtesy of the Department Library Services, American Museum of Natural History.)

MOUNT CARMEL

More important, however, were discoveries made in the Middle East and again in South Africa. The Middle Eastern finds were the result of excavations on the slopes of Mount Carmel in Israel (then Palestine), just south of Tel Aviv and little more than a mile from the shore of the Mediterranean Sea. There, in 1931 and 1932, a joint Anglo-American archaeological expedition discovered the remains of at least a dozen fossil humans in two different caves. The one complete skeleton from the cave of Mugharet et-Tabūn was of a female that was in other respects indistinguishable from the "classic" Neanderthals of Europe. This at least demonstrated that the Neanderthals were not just a limited European phenomenon. This point was also demonstrated by the simultaneous discovery of a series of skulls, approaching Pithecanthropine form but at the edge of the early Neanderthal spectrum, only twenty miles downstream from the site of the original Pithecanthropus discovery at Trinil in Java. After World War II, more full Neanderthals were found at Shanidar cave in northern Iraq, again attesting to Neanderthal distribution.

The second Mount Carmel Cave, Mugharet es-Skhūl, yielded the remains of at least ten individuals, but rather than exhibiting fully Neanderthal form, they displayed characteristics that were halfway between Neanderthal and Modern form. This has earned them the designation of "Neanderthaloid"—that is, recalling the Neanderthals on the one hand, but not to a sufficient degree to separate them from Moderns on the other.

By one of those little ironies that only fate can arrange, the major burden of description and interpretation fell to Sir Arthur Keith. He who had been so critical of Dart's attempt to make sense out of a mixture of simian and human traits was now confronted with the task of interpreting a mixture of Neanderthal and Modern ones, and his vacillations between alternate possibilities satisfied no one. To add further confusion to the matter, opinion tended to regard both Mount Carmel sites as being dated to an interglacial before the onset of the last Pleistocene ice advance. If the date were correct (and, as it turns out, it was not), that would have made them older than the "classic" Neanderthals of Europe. Under such circumstances the form of Tabūn made reasonable sense, but that of Skhūl would represent a dilemma. Keith tried to face the problem squarely, but he could not decide whether Skhūl indicated the hybridization of a fully Neanderthal population (represented by Tabūn) and the long-sought but as-yet-undiscovered Modern one, or whether it was a Neanderthal population in the throes of rapid evolutionary change. He seemed to favor the latter explanation, although others have stressed the former.

In any case, the issue has been drastically changed by recent refinements in dating which have removed both Mount Carmel sites from the last interglacial. It now appears that Tabūn is about 60,000 years old, which makes it a contemporary of full Neanderthals elsewhere. Skhūl is around 35,000 years old, being halfway between proper Neanderthals and accepted Moderns in both appearance and date.

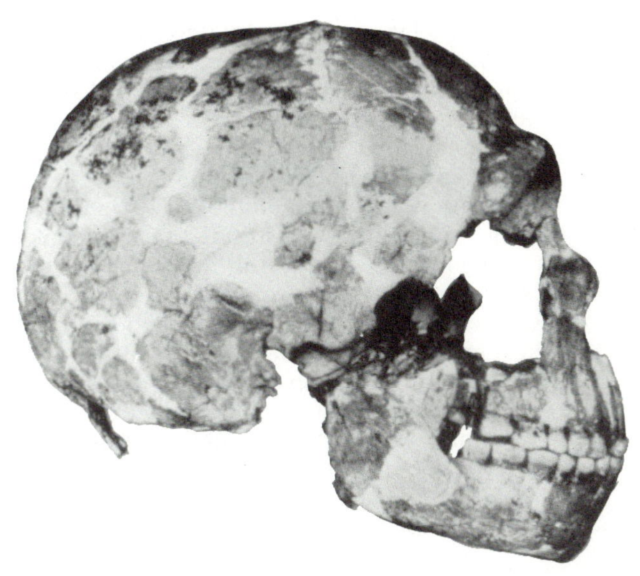

The skull of a classic Neanderthal woman from the cave of et-Tabūn. (Courtesy of the Clarendon Press, Oxford.)

ROBERT BROOM AND MORE AUSTRALOPITHECINES

Because of the dating difficulties, the Mount Carmel problem remained unsolved for a full thirty years, and a vestige of interpretation based on the erroneous earlier date remains with us still. Meanwhile, in 1936, old issues of another sort were reopened, and this time the evidence was sufficient to ward off attempts to sweep them under the rug. Just when it seemed that the furor over Dart's *Australopithecus* had been reduced to a memory, another fortunate explosion occurred. This particular blast took place in a limeworks quarry at Sterkfontein, some miles north of Johannesburg in the Transvaal area of South Africa. Fossil bones were discovered as a result and, by good fortune, were delivered into the hands of the venerable vertebrate paleontologist and physician, Dr. Robert Broom. Fragments of nearly half a dozen creatures were included, among them a complete adult skull. Broom immediately recognized the similarity of these to *Australopithecus*. Eventually he advocated an entire taxonomic subfamily, the *"Australopithecinae,"* to include them all, but he believed that the new finds were different enough to warrant new generic and specific designations. He called them *"Plesianthropus transvaalensis."* Time has shown that they are simply adult versions of *Australopithecus,* and so, for the present at least, we simply refer to them as Australopithecines.

The upright carriage of the head and the extraordinarily human appearance of the distal end of a femur contributed to the suspicion that these creatures may have been erect walking bipeds, and it began to appear as though Dart had not been so rash as his detractors had claimed.

Nor was this all. Two years later, in 1938, on a farm named Kromdraai some two miles from Sterkfontein, more Australopithecine remains were discovered and brought to the attention of Dr. Broom. Fragments of skull, jaw, teeth, arm, and foot bones suggested the presence of a similar but more robust creature, which Broom called *"Paranthropus robustus."* We refer to these as the late or robust Australopithecines.

Now, with the weight of Broom's years of paleontological experience and the quantity of accumulating evidence, the Australopithecines could no longer be passed off as figments of a youthful fancy. However, consistent with the principles of hominid catastrophism, modified or extreme, most anthropologists were convinced that fossils of modern form would yet be discovered in the Early Pleistocene, and they continued to support the view that the Australopithecines were simply another side branch from the main stem of human evolution that had become extinct without issue. To be sure, a few anthropologists—notably the American student of the Neanderthals, Hrdlička, and Franz Weidenreich—complained that the human evolutionary tree portrayed by most scholars was in effect all branches and no trunk, but real qualms did not develop until after World War II.

G. H. R. VON KOENIGSWALD

With the close of the 1930s, war came to the West as it had earlier come to the East, and, as invariably happens under such circumstances, fossil hunting ceased and evolutionary studies were seriously impeded. Fortunately, the conflagration was delayed in Southeast Asia, and work continued for a while in Java. The last major find of that era was made by the Dutch paleontologist G. H. R. von Koenigswald, who had been responsible for further Pithecanthropine finds in Java throughout the late 1930s. Just before the Japanese occupation in 1941, von Koenigswald found a small fragment of a rugged mandible with a few large but clearly human teeth to which he gave the relatively jawbreaking name *"Meganthropus palaeojavanicus."* A similar mandible was found in the same area in 1952, and for a while some anthropologists thought that these belonged to a genuine Australopithecine. This, if true, would be the only example of an Aus-

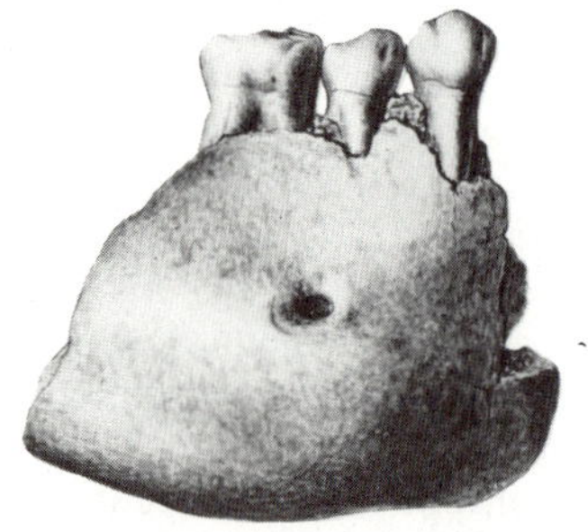

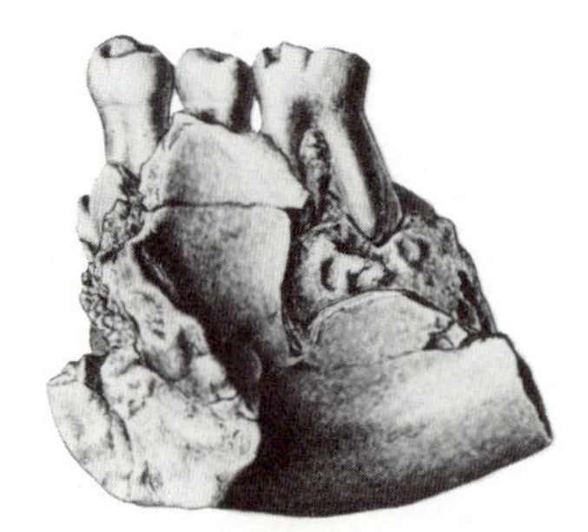

"Meganthropus palaeojavanicus," a robust Pithecanthropine from the early Pleistocene of Java. (Courtesy of Dr. G. H. R. von Koenigswald.)

Dr. Robert Broom (1866–1951), physician and paleontologist, with a cast of Plesianthropus. (Courtesy of the Department Library Services, American Museum of Natural History.)

tralopithecine outside of Africa and therefore of extraordinary significance. The material is really too skimpy for a definitive diagnosis, but, using what is available, comparative analysis shows that it cannot be distinguished from robust Pithecanthropine form. It may be considerably earlier in date than the Middle Pleistocene Pithecanthropines from elsewhere in Java and in China, and its robustness could well be an indication that it is closer in form as well as in time to the Australopithecine ancestors, wherever they might have been.

chapter six

Recent Discoveries

STERKFONTEIN

In 1947, soon after peace made it possible for him to return once again to his prehistoric research, Dr. Broom, now more than 80 years of age but undiminished in energy, resumed his investigations at Sterkfontein. Within a short time Broom discovered quantities of Australopithecine remains, among which was a nearly complete half pelvis. This bone was remarkably like a small Modern human pelvis and, of all the Australopithecine fragments found to date, most clearly demonstrated their erect and bipedal mode of locomotion. Yet traditional anthropological doubts were raised, and it was hinted by some that perhaps the pelvic fragment properly belonged to the long-sought "true man" and had nothing to do with the Australopithecines. Possibly, they hinted, "true man" was hunting the Australopithecines. Variants of this argument continue to be offered concerning other fragments of evidence, but, like the cry of "pathological" that was repeatedly applied to early human fossils in the late nineteenth century, this has begun to sound more than a little strained.

MAKAPANSGAT

In 1947, too, Raymond Dart returned to the arena of Australopithecine research. His work was concentrated on a deserted lime works dump at Makapansgat, some 200 miles northeast of Sterkfontein. This dump produced a wealth of Australopithecine fragments, evidently of the same sort as those of Taung and Sterkfontein, although Dart gave them a new specific name. Within two years Dart had also recovered substantial fragments of pelvis that, if anything, were even more human in appearance than those from Sterkfontein. Also in 1947, Sir Arthur Keith published a handsome and gracious apology, noting that it was he who had been rash and hasty in 1925, and that time and events had proven Dart's interpretation to be much more nearly correct than his own.

SWARTKRANS

Starting in the succeeding year, 1948, Broom and his assistant, J. T. Robinson, began work at another site, Swartkrans, a short distance from Sterkfontein, where they soon found Australopithecines of the same sort that Broom had found at Kromdraai ten years before. Adding still another dimension to the early Pleistocene fossil picture as seen in the Transvaal of South Africa, a pocket in the Swartkrans site produced another form of human fossil. The first fragment, a nearly complete jaw found in 1949, was initially given the name of "*Telanthropus capensis,*" but comparative study convinced Robinson and others that it is not distinguishable from the Pithecanthropines of Asia. As such, it is a find of the greatest significance, and just a hint of what Africa was to produce in the years that followed.

Dart and his assistants at Makapansgat and, following Broom's death in 1951, Robinson at Swartkrans and Sterkfontein, continued to recover fossil fragments of various kinds during the succeeding decade. More recently, after Dart and Robinson retired from active field work, their lead has been followed by C. K. Brain at Swartkrans and Phillip V. Tobias at Sterkfontein. The discoveries have continued to accumulate.

During the decades since the end of World War II, human fossil remains from all corners of the Old World have been discovered that represent all of the stages of human evolution, as well as an increasing fossil record of the prehuman ancestors back some tens of millions of years earlier. Some, such as the Neanderthal discoveries at Shanidar cave in northern Iraq (starting in 1953) and more recently in China, and Pithecanthropine remains from Tautavel (Arago) in the French Pyrenees, Greece, China, and now India, are of major significance, but so far the finds from sub-Saharan Africa have not yielded the center of the stage. The scope of this book does not allow me to indulge in the treatment of more than a fraction of these fascinating finds, although I must confess that my enthusiasm is in no way reduced by finding myself confined to the discoveries of greatest dramatic import.

MARY AND LOUIS LEAKEY AND OLDUVAI GORGE

Certainly one of the most dramatic facets of this historical review is the account of the East African discoveries initiated by the work of the late Dr. L. S. B. Leakey and continued by his wife, Mary Leakey, at Olduvai Gorge, Tanzania (then Tanganyika). Although the Olduvai finds which made them famous did not begin to occur until 1959, the basis that makes these discoveries so important goes back more than a third of a century earlier, to the time when Dr. Leakey first started making expeditions to this area. Leakey's general experience in East African prehistory goes back even further, but he first visited Olduvai Gorge in 1931 and shortly thereafter discovered stone tools of a type as crude as, or cruder than, those of any known tradition. Because of their location, they have been called Oldowan tools. They precede the Acheulean and are now regarded as constituting the oldest tool-making tradition in the world. The Leakeys' continued work in Olduvai Gorge has revealed the development of the

Oldowan tradition through the ascending layers until it becomes the familiar Acheulean "hand axe" tradition recognized a century ago by Boucher de Perthes at Abbeville and widely distributed throughout the Middle Pleistocene of the Old World. Here, then, is the Lower Pleistocene parent of all subsequent human cultural traditions including our own: a confirmation of Darwin's prediction made so long ago!

As if this were not enough of a contribution to have made during a lifetime, the Leakeys' continued persistence in the face of formidable financial and environmental obstacles finally rewarded them with the discovery of the manufacturers of the Oldowan tools—the bones of what must be our own ancestors as well as those of others who may have been collaterals, although there still is a good deal of uncertainty about which falls into what category. On July 17, 1959, the Leakeys discovered their first fossil "man." Initially they followed the annoying paleontological practice of giving it a new generic and specific name before it was systematically compared with anything else. They called it *"Zinjanthropus" boisei,* although it turns out to be nothing less than another Australopithecine. Like the Swartkrans specimens from South Africa, this find differed from many of the other specimens then known by being more robust—so much more robust, in fact, that most specialists now regard the new species name of *boisei* to be fully warranted. This makes the name properly *Australopithecus boisei.*

The discovery of "Zinj," as it continues to be unofficially called, attracted worldwide attention and financial support, and it initiated a new era in paleoanthropology. Within just a few years, the Leakeys had

The "Zinj" skull from Olduvai Gorge, a robust Australopithecine, with an artist's idea of what the mandible should look like. (Photo by R. I. M. Campbell, Nairobi, Kenya.)

Olduvai Gorge, Tanzania. (Courtesy George H. Hansen.)

uncovered the remains of gracile Australopithecines as well as Pithecanthropines at various levels in Olduvai Gorge. Although their work has been of the greatest value to students of human evolution, it has not been an unmixed blessing. Underlying Dr. Leakey's many productive years in the field was the unshaken conviction that nothing so crude as an Australopithecine could possibly have been ancestral to subsequent human forms. He seemed to feel that the only proper ancestor for modern humanity was none other than ancient "true man," and each new discovery was hailed as evidence for this patently nonevolutionary view until proper comparative study showed it to be one or another already known taxon. At one point, Dr. Leakey assembled a mixed collection of Australopithecine and Pithecanthropine material from Olduvai and tried to claim that this, which he called *"Homo habilis,"* was the long-sought "true man" that was ancestral to ourselves.

At the time I wrote the first edition of this book, I had felt that the Australopithecines could be lumped together as a single, if variable, stage in human evolution that evolved as a group into the Pithecanthropines, and that if this were true, then clear representatives of those two distinct stages could not be present at the same point in time. And, indeed, the various forms that Dr. Leakey mixed in this claimed taxon came from very different time levels. Dr. Leakey, however, kept trying to show that this or that new find showed that different forms existed at the same level, always hoping that one might be shown to be his "true man." Because each new piece of evidence always turned out to have some flaw, I became convinced that Dr. Leakey's actions were analogous to those of "the little boy who cried wolf." One of my colleagues reminded me, however, that in the old didactic story, eventually there actually *was* a wolf. And so there was in this case, as we shall see, but the confusions caused by the repeated claims for this, that, or the other status remain with us, and, along with other problems of interpreting the evidence, will continue to plague us for some time to come.

RICHARD LEAKEY AND EAST TURKANA

Dr. Leakey's richly productive life came to an end in 1972, by which time he had seen the fruitful beginnings of the continuation of his work by his son Richard, in a promising area just east of Lake Turkana (then Lake Rudolf), in northern Kenya not far from the Ethiopian border. At that time, Richard Leakey's field crew had found a skull that was not quite a Pithecanthropine but was certainly tending in that direction. This, the cranium ER-1470, was half again as large as "Zinj" and nowhere near as heavily buttressed and flanged, but it was of roughly the same date. Clearly the transition that was to produce the Pithecanthropines from the Australopithecines was half accomplished. The face was still large and of Australopithecine form, but the braincase was definitely modified in a Pithecanthropine direction, both in its external and its internal features. Unfortunately, it too was given the label *Homo "habilis"* so that now this "category" includes some specimens that are clearly Australopithecine, some that are Pithecanthropine, and ER-1470, which is clearly in between. Actually, the difficulty in assigning ER-1470 to a known category points up just how arbitrary it is to divide an evolutionary continuum into pat stages, however useful these may be in general.

But if one group of Australopithecines evolved into Pithecanthropines, another group clearly did not. Both forms are now shown to be

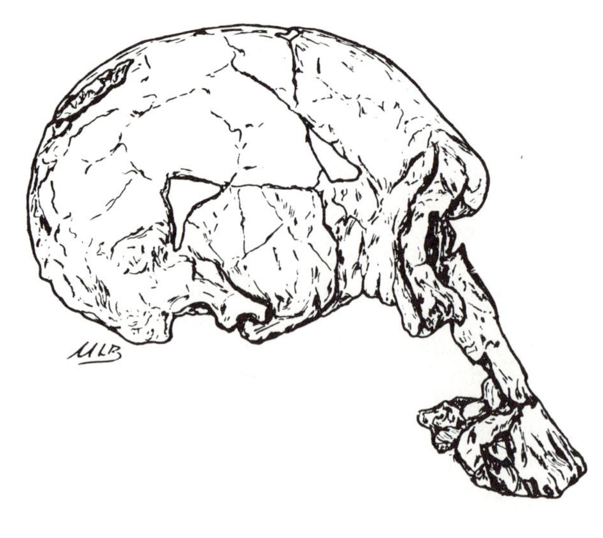

KNM ER-1470, a transitional Australopithecine/Pithecanthropine skull from east of Lake Turkana. (Drawn from a cast.)

contemporaries in the East Turkana area being worked by Richard Leakey and his associates. The skull ER-406, found at Ileret in 1969, is an obvious robust Australopithecine (brain size = 510 cc.), and yet it is a contemporary of the 1.3 million-year-old ER-3733 skull found at Koobi Fora in 1975 (also in the East Turkana area), which is an unmistakable Pithecanthropine (brain size = 800–900 cc.). It was the discovery of ER-3733 that really provided the clincher to the view that some form of *Homo* lived at the same time as some form of *Australopithecus*, and one could argue that this is as close to the real "wolf" as we are likely to come.

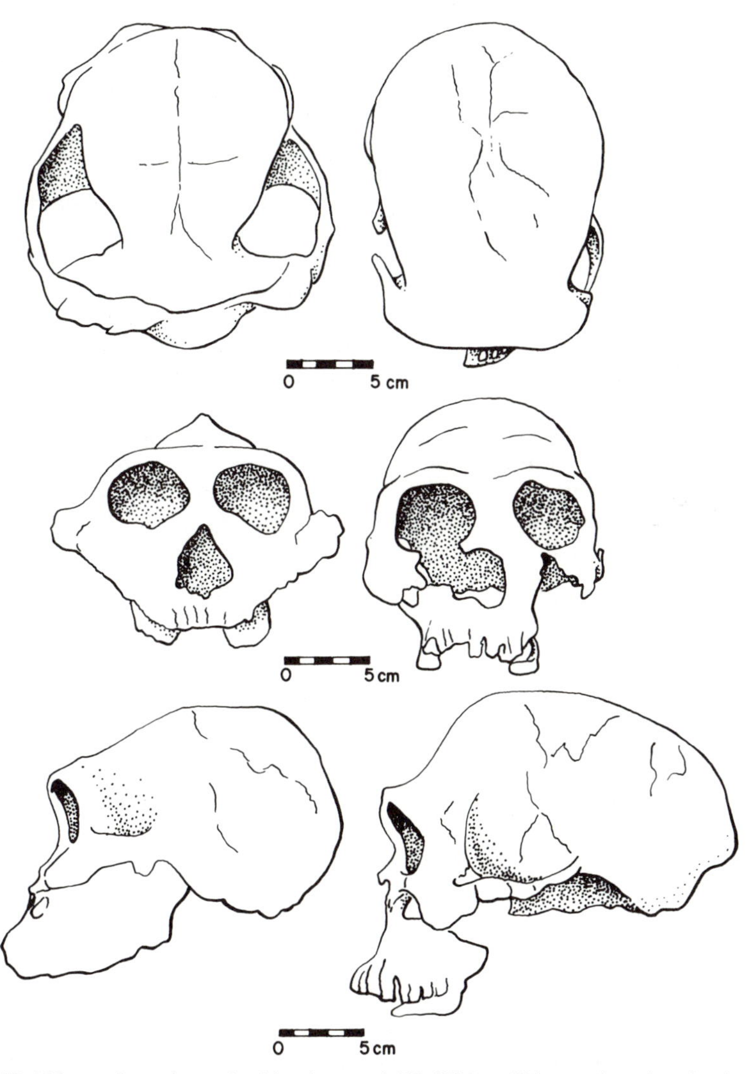

ER-406, a robust Australopithecine, and ER-3733, a Pithecanthropine, both contemporaries in the East Turkana area of Kenya about 1.3 million years ago.

DON JOHANSON AND ETHIOPIA

The continuing work of the Leakey family during the last few years has given us the pieces with which we can provide a rationale for how the Australopithecines ended and how the Pithecanthropines began. The same years have also produced evidence that helps us understand something of how the Australopithecines themselves began. Starting in 1972, a series of field seasons by a French-American group made significant discoveries in the Pliocene deposits in the badlands of the Afar Depression north of Addis Ababa in northern Ethiopia.

Conditions for pursuing research were considerably less than promising. The climate is harsh in the extreme: an unrelieved, blazing desert heat. And however unprepossessing it may seem, the area has been politically contended for years with intermittent sniping, raiding, armed forays, and out-and-out civil war. But the Pliocene strata lie flat and workable, filled with fossil bones, and datable by the K/Ar technique. So with ingenuity, persistence, and more than a little courage, the French geologist Maurice Taieb and the American anthropologist Donald C. Johanson led crews on yearly efforts until full-scale civil war drove them out in 1977.

Their fieldwork was crowned with abundant success. Their most spectacular discoveries were a nearly complete Australopithecine skeleton late in 1974 and a group of over a dozen more Australopithecines at the end of the 1975 field season. They christened that first relatively complete specimen "Lucy," who turns out to be an interesting specimen indeed. Lucy was barely 3½ feet tall and fully bipedal, although her arms were relatively long in proportion to the length of her legs. Her jaws and teeth, like those of other specimens found there, display a series of features that are intermediate between an ape-like and a human condition. With a date of over 3 million years, Lucy and her companions represent the earliest Australopithecines so far discovered. Given that degree of antiquity, a mixture of human, ape-like, and transitional features is just what we would expect to find. Just to ensure that the collection will continue to be the focus of professional academic controversy, the specimens have been given yet another specific name—*Australopithecus "afarensis"*—and the claim has been put forth that these, and not the South African or Olduvai Australopithecines, are the "true" ancestors of later human populations.

CONTINUING FIELDWORK

These discoveries have led to further efforts in the field, and these efforts have been crowned by further success. In spite of the problems caused by nationalism, professional jealousies, and limited funds for fieldwork, hardly a year goes by without significant new discoveries. Because the early evidence for our evolutionary line is something that appeals to the imagination of the public at large, it is considered newsworthy, and each new find is reported in the popular press.

Unfortunately, while the press generally sees to it that those who are chosen to report on matters of art, poetry, literature, history, politics, and the like actually know something about the fields on which they write, the same is not generally true for science. As a result, many of the news reports dealing with paleoanthropological discoveries display that mixture of misunderstanding and sensationalism that appears so regularly in the media when the topic is science. The headline "New Find Overturns All Previous Theories of Human Evolution" appears every few years in newspapers all over the world. One appeared within a week of the completion of this manuscript in the summer of 1986.

The discovery being hailed was an *Australopithecus boisei* specimen, WT 17000, found by Professor Alan Walker of Johns Hopkins University, a long-time collaborator with Richard Leakey. The find was actually made in August of 1985 southwest of Lake Turkana in northern Kenya, and it came from deposits 2.5 million years old. That is a full million years older than the previously known *boisei* specimens. The brain case is even smaller, and other aspects are more primitive. It is an important and dramatic find, and, in extending our knowledge in significant ways, it has raised a number of questions that could not even have been asked before. But that is how science works, and it hardly constitutes the "overthrow" of all previous theories of human evolution.

Barely a year later, in the summer of 1986, Professor Tim White of the University of California at Berkeley was surveying a portion of Olduvai Gorge in a collaborative project with Don Johanson, now director of the Institute of Human Origins in Berkeley, California. On July 21, Tim White, and subsequently others, found fragments of a skeleton that rivals Lucy in interest. Given the fragmentary nature of fossils that old—and this was found in Bed I, which makes it in the million-and-a-half year range—it is rare indeed to find pieces of *both* the upper and lower limbs of the same individual. Their discovery had pieces of both arm and leg bones, and the preliminary indications are that they have found an Australopithecine that is even smaller than Lucy, but which resembles her in having arms that are relatively large in proportion to leg size. Evidently the Australopithecines, although well-adapted terrestrial bipeds, also engaged in activities that preserved relatively robust arms, and it has been suggested by some that they may have continued to use trees as a place of refuge at night or in times of stress.

As of this writing, Tim White and Don Johanson's new find has not yet come to the attention of the headline writers, but by the time this edition gets into print, there will surely have been a flurry of announcements with varying permutations of the phrase "Revolutionary New Find. . ." But we have spent enough time recounting what was found, when, and by whom, and what they have said about it, and it is time now to get on to an assessment of what it all means.

chapter seven

Evolutionary Principles

Many approaches to the study of human evolution are restricted to descriptions of the fossil evidence, or perhaps to the events surrounding the discovery of the major fragments. However fascinating these may be (and here I suspect my own enthusiasm for the bony details of long defunct hominids may somewhat outrun that of the beginning student or general reader), they do not automatically ensure the full understanding of what is being described. Somewhere along the line, one should encounter a summary of the major principles of evolution in general, and the forces which act on humans in particular. Finally, these should be specifically applied to the human fossil record, and the role that they have played in the production of the specific noted changes should be delineated. This last is the subject of the final chapters; meanwhile, we briefly consider the major evolutionary principles that are important to take into account in human evolution.

TYPES VERSUS POPULATIONS

The study of human evolution starts with the discovery of the fossil evidence. This is then arranged in temporal sequence, and the patterns that can be discovered are interpreted according to standard evolutionary principles. Although nearly all of the current treatments discuss both evolutionary principles and the fossil evidence, one often gets the impression that the actual interpretations are being guided by unrecognized, or at least unstated, sets of assumptions. To the average biologist trained in the Darwinian tradition, the student of human evolution has often appeared to be marching to the beat of a different drummer. Recently this has been made explicit in some treatments of paleontology, and it has been particularly apparent in current paleoanthropology.

We have seen how the lingering tradition of catastrophism has continued to be influential. Some other, equally important vestiges of traditional Western world views have continued to survive as well. One of these is the notion of the "typical" or "ideal." Renaissance artists portrayed human form according to what they felt were ideal proportions, and post-Renaissance biologists stocked their museums with what they hoped were typical representatives of the plants and animals of the world. In Platonic philosophy, the essence of reality was a perfect idea that was reflected in less-than-perfect forms visible in the material world.

Dating from the time of Augustine in the fourth century A.D., currents of Platonic thought, really Neoplatonism, were combined with Christian faith to shape the way people thought about things. In this amalgamated intellectual tradition, the essence of reality, as it had been to Plato, was a perfect idea, but in this case it was an Idea in the Mind of God. Visible form can approach, but it cannot attain, that perfection, for to do so would be to unite it with divinity itself.

As the Renaissance age of exploration and discovery revealed to Europeans the enormous variety of plants and animals that they assumed reflected the manifold ideas in the mind of God, it became a kind of act of piety to list and describe them. To a considerable extent, this provided the impetus for the systematic pursuit of science. Whether it was Kepler working out the laws of planetary motion or Linnaeus identifying the plants and animals of the world, scientists believed they were discovering the dimensions of the mind of God and were thereby coming to understand God's plan for the way things ought to be.

Because all of these categories were thought to derive from a perfect, eternal, and therefore changeless Deity, the assumption followed that these ideal categories were fixed, perfect, and eternal. The ideas of ash tree and aster, lobster and lion, man, mouse, and all the rest had a reality that not only derived from, but proved the existence of, God. To suggest that such might not be the case was tantamount to atheism or blasphemy or worse. Obviously, when Darwin developed his demonstration that the only constant was change itself, operating in response to the interplay of the mechanics of impersonal natural forces, he was perceived by the pious traditionalists as an enemy to the assumed perfection and fixity of their revealed religion.

Darwin's focus on change not only cast doubt on the validity of fixed and changeless types, but it also identified the source of potential change in the naturally occurring variation visible among the individuals of any given species. Individual differences, then, were not just imperfect renderings of an intended ideal type. Instead, they provided a vital demonstration of how species can adapt to changing circumstances.

As a result of Darwin's insights, evolutionary biologists now realize that it is at least as important to know the naturally occurring range of variation of a given group as it is to know its idealized average. Consequently, the typological thinking of pre-Darwinian science has tended to be replaced by the populational approach of modern biology.

Old habits die hard, however, and, although it is often denounced in theory and unrecognized in fact, typological thinking is alive and well and continuing to flourish in the last quarter of the twentieth century. To an extent it is unavoidable in practice. Just to talk about things we have to give them names, and because of the nature of the fossil record, single specimens, often quite incomplete, have to do duty for whole populations and long periods of time. Scholars gain recognition for key discoveries, and inevitably the names and descriptions they give acquire an importance that smacks of typological essence with overtones of sanctity.

Finally, as was noted in the chapter on hominid catastrophism, the field of paleontology in general and paleoanthropology in particular has strong roots in the scholarship of France late in the nineteenth and early in the twentieth century. The French not only rejected a Darwinian view of evolution with its emphasis on populational thinking and the action of natural forces, but stressed a kind of Platonic essentialism and generally avoided a consideration of natural mechanisms at all. The focus on the intricacies and nuances of form without any particular regard for function derives from a tacit assumption that detailed description can reveal a kind of teleological intent. Inevitably, a substantial portion of the typological approach survives in practice in modern anthropology. Recently there has been an increasing effort to provide some kind of theoretical justification, even though this runs counter to the expectations of Darwinian biology.

HUMANS IN NATURE

Even though this book is about human evolution, it takes it as factually demonstrable that human beings are a part of the animal world. Certainly to claim the contrary would qualify as irrational at best. The patterning of our limbs and bodies, the numbers of bones and segments in our appendages, and the way our joints work are repeated down to the last detail in millions upon millions of other accepted members of the animal world. The morphology of the neurons and axons of our nervous system, the chemistry of its synapses, and the transmission of its nerve impulses are absolutely identical. So too are the structural, functional, and chemical details of our skeletomuscular systems, our sensory organs, our circulatory systems, and the whole rest of our basic biological construction. To be sure, we occasionally refer to those whom we consider to be uncommonly inert as "vegetables," but even as we use such a term of opprobrium we realize that the objects of our scorn are not really members of the Plant Kingdom. And yet we do share things with plants that also demonstrate a kinship, albeit more remote, that could not have been the result of accidental independent acquisition. The nucleate cell is the basic building block for plants and animals. The mechanics of cell division, the phenomenon of sexual reproduction, the control of heredity by DNA, and the roles of structural and enzymatic proteins are all essentially the same. We know vastly more about the details of these similarities between living organisms than Darwin did when he realized that a theory of descent from a common origin was the most likely way to account for all of this, and everything we have learned since has simply bolstered the credibility of his insight.

DISTINCTIONS

It is also obvious that there are aspects of difference as well as of similarity, and we should be able to use the enumerated differences as an indicator of time since that point of common origin. In both theory and practice, matters are considerably more complex than that, but the overall generalization remains true. Studying similarities and differences, then, can enable us to arrange the living creatures of the world into various inclusive and exclusive categories that reflect their relationships in terms of relative recency of descent from common ancestors.

We recognize, for example, that house cats and lions have more in common with each other than do cats and dogs. Cats and dogs, however, have more in common than either has with horses or cows, and cats and cows share more with each other than either does with frogs or fish—and so on. By systematic means, we can build up a classification of living things that should reflect their evolutionary history. Most biologists assume, then, that classification is a means of depicting the results of evolution.

But this was not always the case. Many of the traditions that have shaped the practice of classification date from an earlier time, when it was assumed that its object was to demonstrate the categories and logic of God's created world. Of course, people have been classifying things since the dawn of recorded history, and considering the elaborate accuracy of the classifications of nature by nonliterate people from Australia and New Guinea to the Arctic it is certain that people have been engaged in classification at least since the beginnings of language itself.

LOGIC AND NATURE

The formal written traditions of our classifications of the natural world go back to Aristotle. Both his application and his logic were adopted by the scholars of the Medieval Christian church. This, in turn, was adopted by Linnaeus, the eighteenth-century Swedish botanist whose system is universally used today. Our terms Class, Order, Family, Genus, and Species were all used by Linnaeus, but these were, in fact, the very terms of Medieval Aristotelian logic. Medieval scholars assumed that God's world was rigidly logical and that all of its categories were equivalent. These, in effect, were fixed and unchangeable essences determined by God's intent. Classification, then, could be accomplished by strictly logical means.

In practice, many biologists realized that the higher categories were determined by human decisions for our own convenience. Darwin even went so far as to recognize that species themselves, the basic building blocks of any classification, not only had an element of the subjective in their establishment but were also capable of changing through time. This not only was profoundly upsetting to many of Darwin's contemporaries, but has also continued to be a sore subject for many biologists right up to the present day. "If species do not exist at all, . . . how can they vary?" was the anguished objection of one of Darwin's eminent contemporaries. And today, a powerful school of thought has sought to reject this aspect of Darwin's insight and to return to the certainties of the Middle Ages, where species were considered to be fixed, and their relationships could be determined by the strict application of Aristotelian logic.

LOGIC AND COMPUTERS

It is no accident that the emergence of this modern movement coincides with the growth in use of that extraordinarily powerful and useful instrument, the digital computer. The computer is based on elementary certainties: whether a datum exists or it does not, something is true or it is not true, an answer is yes or it is no. Inevitably, an instrument that is so enormously convenient will lead its users to take maximum advantage of its capabilities, but there is a sometimes unrealized consequence. It just may seduce its users into structuring their expectations and the way they handle their data and their problems so that they will be compatible with what the computer does best. The result is that the computer—admittedly not by anything one could call "intent" on its part—may influence the way we think about things. It is just possible that the recent enthusiasm for using a rigid deductive logic to solve the problems of evolutionary relationships is an unconscious byproduct.

THE CATEGORIES OF CLASSIFICATION

Ever since the Middle Ages, there have been repeated attempts to arrange the elements of the living world in a hierarchy that has a completely logical structure. The Linnaean classification of the eighteenth century, now universally accepted, is one such attempt. Plants and animals are put into separate Kingdoms. Within each of these, further distinctions are made at the Phylum level. For example, humans are placed in the Phylum distinguished by the possession of a spinal chord located in the back: Phylum **Chordata.** Fish, reptiles, amphibians, birds, and mammals are distinguished at the Class level. Within the Class **Mammalia,** insectivores, bats, carnivores, rodents, elephants, and many others are distinguished at the Order level. We belong in the Order **Primates** along with lemurs, lorises, tarsiers, monkeys, and apes. We are separated from them at the Family level, where we are noted as belonging to Family **Hominidae.** The great apes belong to Family **Pongidae,** and the close affinity of humans and apes is recognized at what is called the Superfamily level, where we are included as Superfamily **Hominoidea.** In an informal sense this is

The classification of *Homo sapiens.*

Kingdom	**Animalia**
Phylum	**Chordata**
Class	**Mammalia**
Order	**Primates**
Family	**Hominidae**
Genus	***Homo***
species	***sapiens***

recognized by the term "hominoid." The familial distinctions are informally recognized by the use of the terms "pongid" for apes and "hominid" for humans. Genus *Homo* and species *sapiens* complete the full formal classification from our own anthropocentric point of view.

All of these major levels of classification can be lumped at the "super" level or divided at the "sub" level and further divided at the "infra" level, with more attention paid to the dividing than to the lumping. So we have **Suborders, Infraorders, Superfamilies, Subspecies,** and so on. The professional arguments over which is included with what, and what is distinguished from which, are endless. Over two hundred years ago, the French biologist Buffon declared that it was all arbitrary, since classification was practiced by human beings for their own purposes, and none of the categories had any logically independent reality. Eventually, he admitted the separate reality of species, but not of the higher categories. At an operational level, this has been the position of most biologists ever since.

SPECIES AND REALITY

But even at the species level, none other than Darwin himself noted that categorization had an element of the arbitrary. Many have found this a profoundly disturbing point and have refused to accept it. The arguments on that issue have raged back and forth, and recently there has been a resurgence of the stance that species do have a categorical reality. In part this may be a reflection of the human desire for elementary certainty akin to the Medieval faith in the verity of the units of God's created world, and in part this may be a result of the pragmatic utility inherent in being able to take for granted the reality of the units the computer allows us to manipulate with such ease. And whether or not species are regarded as validly definable and strictly comparable units, virtually all modern biologists use them as though they were.

SPECIES AND CHANGE

That species have descended from previous species is a fact upon which all are in agreement. How this occurred is a matter of profound and continuing disagreement. The source of that disagreement is based partly on the nature of the species with which the various investigators are most familiar, and partly on the different intellectual traditions in which those investigators were trained. This book is concerned with change in the human species, so the choice of interpretation in the ongoing arguments about the nature of specific change is obviously conditioned by a familiarity with the human biological evidence, present and past. The intellectual tradition on which the interpretation of that evidence is based is that which was established with the publication of Darwin's *Origin* and which has grown with the addition of a genetic—especially a molecular genetic—perspective.

Given this basis, it is perfectly clear that change in the human line of development has been a slow, gradual phenomenon. This does not mean that it has always proceeded at the same rate, or that the traits that can be assessed were all undergoing change simultaneously. But even though this may be "perfectly clear" if looked at in this way, it is not the view of the majority who work in paleoanthopology. Why this should be so is a demonstration of the strength and continuity of the essentialist tradition stemming from Medieval Neoplatonism.

CLADES AND CLADISTICS

This tradition is currently known by the name of "cladistics," or, as its proponents prefer, "phylogenetic systematics." A clade is simply the continuation of a given lineage or "monophyletic unit" through time, and cladistics is the use of particular kinds of traits for the formal assessment of the relationships and distinctions of clades. The results can be presented in a cladogram.

For example, the position of humans in relation to their closest Primate relatives can be clearly seen in the cladogram presented here. There is no assumption of length of time or rate of change in a cladogram, and there is no identification of a common ancestor. Relationship is based on the recency of the split in clades from an assumed common ancestor and is expressed in the sharing of recently evolved or "derived" traits—"apomorphies," to use the formal term. Similarities based on the retention of common "primitive" traits—"plesiomorphies"—cannot show closeness of relationship. The hitch in this procedure is in determining which traits are primitive and which are derived. Another hitch can occur when what appear to be similar and recently derived traits in different organisms came about by "convergence" from somewhat different previous states.

Careful consideration and skeptical caution can overcome some of these problems and assure that a cladistic approach can yield valuable information on organic relationships. However, another assumption made by many practicing cladists cannot be dealt with so easily. As the field is

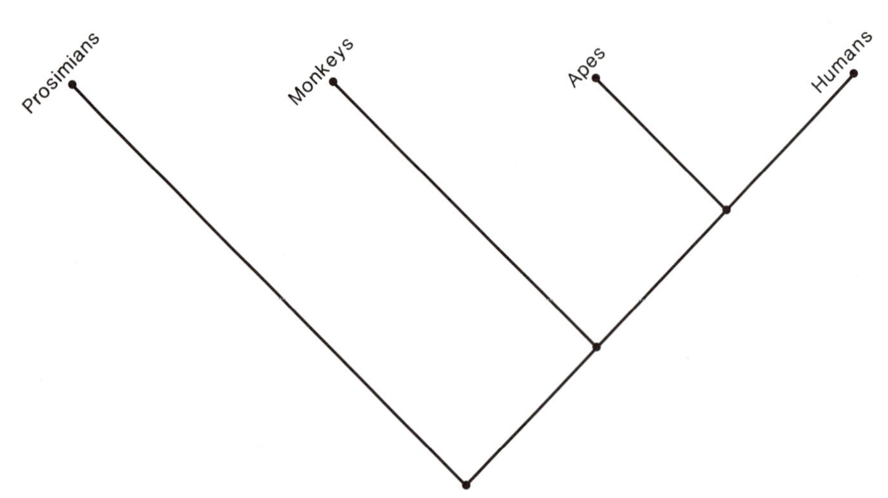

A cladogram showing the relationships of the major groups of primates in terms of the relative recency of splitting from common ancestors.

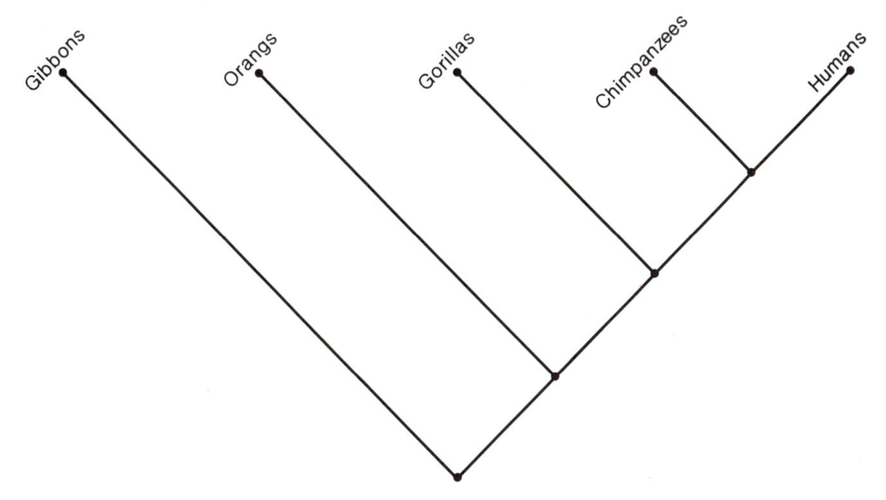

This cladogram shows the relationships between humans and the various members of the Ape Grade of Primates, based on a combination of anatomical and DNA similarities and differences.

constituted at present, it considers it to be axiomatic that a species *cannot* change gradually through time to the extent that it has to be granted new specific status. Evolutionary changes are assumed to occur only as a result of "speciation events" that are represented by the nodes or branch points of a cladogram. At this point, the parent species presumably ceases to exist, and two "sister" species of equivalent rank arise. Each of these then continues without change either to extinction or to a new branching point, at which two new sister species arise.

As a presentation of the nature of evolutionary change, this represents an assumption—a priori—that is simply taken on faith. In practice, there is no consideration of evolutionary process or the interplay of selective force and adaptive response. The result is remarkably like the picture presented by Cuvier's catastrophism early in the nineteenth century, which regards change as occurring suddenly, for undiscoverable reasons, and away from the region under examination. The new form, which spreads by migration, then prevails until the next sudden change.

PUNCTUATED EQUILIBRIA

Closely related to cladistics but differing in some details is the theory of punctuated equilibria. This maintains that the normal state of affairs is the persistence of species unchanged throughout the duration of their existence, a phenomenon referred to as "stasis." Gradual response to selective force is generally denied, and change, when it occurs, takes place suddenly at the isolated peripheries of the species range. The successful new species that results from this then persists in unchanging equilibrium until the next punctuation.

Just as do orthodox cladists, punctuationists regard evolution as a thing of fits and starts with long periods of stasis in between. Unlike cladists, they do not insist upon the simultaneous emergence of two sister species and the disappearance of the parent species during the speciation event. A parent species can persist even after giving rise to a daughter species, and may even outlive it.

Some advocates of punctuated equilibria are not cladists in the strict sense, then, but all cladists view the picture of organic life through time as one of long periods characterized by the unchanging persistence of living forms, punctuated at intervals by sudden change. Inevitably, when dealing with the human fossil record, both cladists and punctuationists prefer the traditional views previously described as hominid catastrophism. Adaptation is ignored, and extinctions and invasions are postulated, even though the archaeological and fossil record does not provide convincing support.

Needless to say, this is not the orientation of this book. But that does not mean that the stance taken by the punctuationists and the cladists is always wrong. It is just not always right. We have to find out from the evidence itself whether and when stasis is the case, where speciation takes place, whether it is the result of lineage splitting, and if adaptive response to selective force change can indeed account for gradual change through time. As we shall see, all of these phenomena can be observed at one time or another in the human fossil record.

GRADES

While clades represent lineages continuing through time, there is another way of looking at organic forms in comparative perspective, and this is the consideration of grades. Grades imply comparable levels of adaptive integration where similar traits allow the pursuit of similar life ways. Birds and bats represent a flying vertebrate grade, although they are not linearly related.

Actually, bats themselves are an interesting example of the grade of flying mammal since they are not all that closely related to each other. The insect-eating bats are really airborne insectivores, but the fruit bats or "flying foxes" may be more closely related to true Primates—a kind of lemur that has independently acquired the powers of flight.

Within our own order, Primates, there are four recognized grades—or five if the fruit bats are considered to count. The first is the Prosimian Grade, which includes the lemurs of Madagascar, the galagos of Africa, the lorises of Africa, India, and Southeast Asia, and perhaps the Tarsiers of Indonesia and the Philippines. Prosimians retain many of the traits of their nonprimate ancestors. They possess sensory whiskers and a relatively acute sense of smell. Their vision tends to be less than fully binocular and stereoscopic, and their brain size and manipulative capabilities are less well developed than those in the other primate grades. But they are fully arboreal and can be considered slightly modified survivors of the first mammals that took to the trees and became primates.

There next is the Monkey Grade. Like the Prosimians, monkeys are basically arboreal quadrupeds, although some, such as baboons, have become secondarily terrestrial. Fully stereoscopic binocular vision has been achieved, the sense of smell is less important, brain size and manipulative capacity have increased, and the basic dietary focus is on the fruit, berries, nuts, and leaves that the tropical forest canopy produces in such abundance.

Beyond the Monkey Grade, is the Ape Grade. The tendency for relative cerebral expansion is carried yet another step, and the normal primate tail disappears altogether. At the same time, there is a general average increase in body size associated with a different habitual mode of locomotion. Instead of being an arboreal version of a terrestrial quadruped, full-scale quadrumanous (four-handed) climbing or clambering is the characteristic means of getting about. Arm swinging and hand-over-hand progression beneath the branches rather than quadrupedal running on top of them becomes more common. This is what allows for the development of a larger body size without limiting the apes to only the sturdier limbs and tree trunks; it also improves the chances for access to the fruits and related edibles that grow at the tips of the smaller branches.

The final grade is the Human Grade, which is characterized by a still further expansion of the brain, by an erect-walking bipedal mode of locomotion, and by the development of culture as a principal adaptation. Within the Human Grade are four grade-like levels of adaptation that, for the purposes of this book, I have termed the Stages of Human Evolution. These are the Australopithecine, the Pithecanthropine, the Neanderthal, and the Modern Stages; each will be treated in its own chapter.

NATURAL SELECTION

The first and most important of the evolutionary principles is encompassed by the term "natural selection." Like many other basic principles in other areas of thought (culture for the anthropologist or entropy for the physicist, for example), natural selection is hard to define concisely, although natural scientists are virtually unanimous in its usage. Realizing that this is an oversimplification, we can regard natural selection as being the sum total of naturally occurring forces that influence the relative chances for survival and perpetuation of the various manifestations of organic life.

Credit can be given to Charles Darwin for using this as the major explanatory principle necessary to account for the cumulative change apparent in the history of any given organic line. Occasionally the principle has been tersely expressed as "survival of the fittest," although it was not Darwin who phrased it that way, and it has been justly criticized as being not quite accurate. In the recent past it has been pointed out that evolutionary survival is determined more by reproductive success than by physical strength. The suggestion has been made that the phrase might be modified to read "the survival of the fit." Actually, a certain amount of verbal quibbling must inevitably surround any attempt to produce a precise definition because, if fitness is described in terms of the production of viable offspring, then the fittest will obviously be those who produce the most and whose traits will be most frequently represented in subsequent generations.

As environmental forces change over time, the characteristics that have greatest survival will not be those which were most valuable at an earlier age. However, in order that environmental forces may effect a change in the characteristic appearance of the species in question—that is, in order for natural selection to produce evolution—some source for the new traits must be postulated. In Darwin's day this source was unknown. On the basis of extensive observational experience, however, Darwin knew that variants were always being produced, so he simply accepted their existence "on faith," without knowing where they came from. Some of the most bitter attacks against him came from those who recognized that he did take it that way.

Today we recognize that the faith Darwin had in his observation that variation occurs in the normal course of events was faith well grounded. When developments in the field of genetics led to the recognition of mutations as the source of variation, it was soon explicitly realized that "mutations provide the raw material for natural selection." As has been mentioned, this was the point where evolutionary thought and genetics joined to produce what has come to be called "the synthetic theory of evolution."

THE PROBABLE MUTATION EFFECT

If the summed forces of nature working on organic variability can be regarded as the most important principle in the production of evolutionary change, then the next most important dimension must be that which is determined by the nature and frequency of the sources of variation themselves. Simply stated, the nature of mutations, their frequency of occurrence, and the probable effect they have are of an importance second only to natural selection.

Although a discussion of genetics at the molecular level may seem rather a long way from the human fossil record, its importance will become clear during the treatment of all but one of the major changes that characterize the course of human evolution. In essence, the story goes like this: Recent research has identified the basic genetic material, demonstrated its structure, and suggested the mode of action whereby it controls organic form and function, as well as replicates itself. To say that the basic genetic material is DNA is true enough, but to say that a mutation is an error in the attempt by a DNA molecule to copy itself, though true, is not precise enough for our purposes. To appreciate the significance of the average mutation, we must first have some idea of how genetic control normally works.

For the moment we are concerned with two kinds of organic molecules: nucleic acids and proteins. Both are polymers—that is, they are chain-like structures whose links are called amino acids (in the case of proteins) and nucleotides (in the case of nucleic acids). The full nucleic acid molecule is a double chain composed of complementary or mirror-image halves. However while the double chain of a nucleic acid is built up of only four different kinds of its basic building blocks, the nucleotides, the single chain protein molecules have more than twenty kinds of amino acids available for their construction.

Both kinds of molecules are of vital importance. Proteins not only form the structural components of which living organisms are constructed (bone, muscle, fiber), but they also constitute the organic catalysts called enzymes, hormones, and other such enabling molecules as hemoglobin, insulin, and adrenaline, without which normal metabolic functioning and growth could not occur. DNA, on the other hand, mostly remains within the nucleus of the cell, where it provides a source of information for the construction of protein molecules.

At first it was not known how the four nucleotides of nucleic acids related to the twenty plus amino acids of proteins. To simplify the rather complex process that subsequent research has shown to be involved, let us put it this way: Various sequences of nucleotides, taken three at a time, can specify (serve as the code for) given amino acids. The nucleotides, with the aid of specific enzymes, fasten together the amino acids by means of a phosphate energy bond. In the course of protein production, quantities

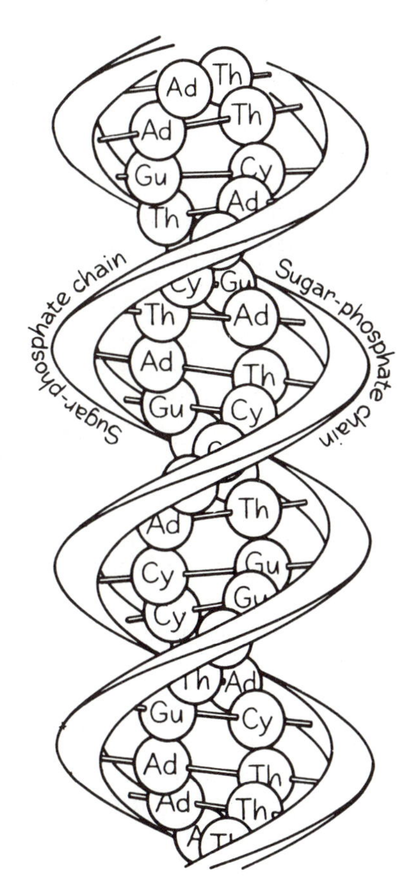

The double helix, or Watson-Crick, model of a DNA molecule.

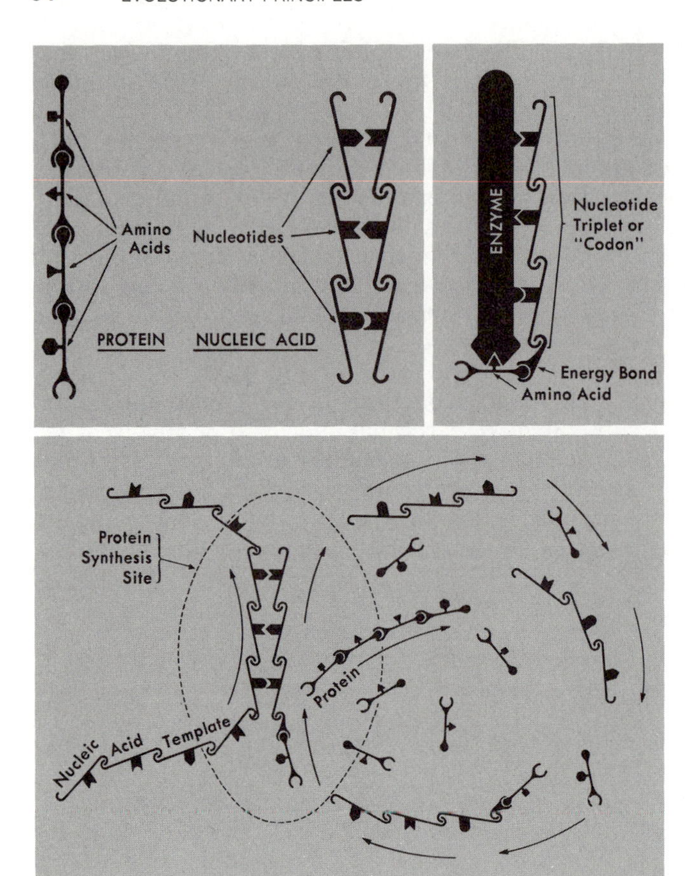

Schematic representation of a protein molecule, made up of amino acid units in a chain; and of a nucleic acid molecule, made up of nucleotide units in a chain. Right, attachment of a specific amino acid to a particular nucleotide triplet with the aid of an enzyme specially constructed for that purpose. Bottom, amino acids being towed by nucleotide triplets to the site where they are hooked together to form proteins.

of nucleotide triplets attach themselves to free amino acids and tow them to the sites of protein synthesis within the cell, where they are lined up and snapped together.

This brief and grossly oversimplified description is not intended to be complete. The point to be made is that a sequence of three specific nucleotides identifies one particular amino acid. If by chance an error is made in the replication of the nucleic acid molecule, the smallest identifiable change will be the modification of a single nucleotide. Although a single nucleotide does not correspond to an entire amino acid, the modification of any one nucleotide will indeed change the nature of the triplet to which it belongs, and in all likelihood this will code a different amino acid from what had previously been the case—if indeed it codes any.

Two of the most likely changes at the single nucleotide level result in still further complications. The addition or deletion of a nucleotide will not only modify the triplet within which it occurs, but will also change the nature of the triplets that follow. Obviously, this means that more than one amino acid in the related protein will be changed, and it seems quite clear that the protein will not work in the way it was intended. A change in even one amino acid can drastically alter the function of the protein of which it is a part, as can be seen in sickle cell anemia and a variety of other deficiency diseases of an inherited nature.

What we have been discussing forms the basis for the fact realized half a century ago: that the great majority of mutations will be disadvantageous to the organism in which they occur. Actually this statement, though correct, assumes a static picture in which environmental forces remain unchanged—that is, it assumes that any modification in an organism will be detrimental to it. To be sure, when an organism is well adapted to its environment, most alterations in its form that arise by chance will not be advantageous. But the environment itself may undergo a change, in which instance some structures may be less important for the organism's survival. Although this is not a major problem in evolution, it is one that continually arises, and, because it has apparently occurred in the course of human development, it deserves some attention.

When some major change occurs in the environment, or when a creature enters an entirely new environment, it may suddenly find that it possesses some structures which, while not disadvantageous, are of no particular value to it. Because the structures in question are neither selected *for* nor selected *against,* they are at the mercy of random modifications in the genetic material that controls their appearance. It is now pertinent to ask what sort of variations are likely to occur. If we are dealing with a feature of gross morphology (teeth, horns, pigmentation), we must realize that these are the results of a period of growth during which development is influenced by the sequential interaction of a sizable series of enzymes. Since enzymes are protein molecules, evidently many of them will be subject to direct genetic control, with mutations affecting their amino acid constituents.

Random variation in a morphological feature usually occurs in modifications of this growth process, and the most common modification of the growth process is in the form of mutations affecting the controlling enzymes. At this level, the expected change is the alteration of a single amino acid, and almost invariably the result is either that the enzyme fails to work altogether, or that it does not work as well as it did in its unaltered form. If a growth enzyme fails to work, or works only to a reduced degree, then the structure that depends upon it either will fail to occur, or else will occur in reduced or only partially developed form. Stated briefly, the most likely effect of the most likely mutation will be the reduction of a structure which depends upon it—that is, the result of the probable mutation effect is structural reduction. In a normally adapted organism, such reductions will, of course, be disadvantageous. On the other hand, whenever circumstances are altered so that a structure no longer has the same importance for survival that it previously had, we can predict that the probable mutation effect will produce its ultimate reduction.

Normal................DAD AND NAN CAN ADD DAN AND ANN

Mutated by (D)
addition of D.......DDA DAN DNA NCA NAD DDA NAN DAN N
or, (A)
deletion of A.......DDA NDN ANC ANA DDD ANA NDA NN
or,
substitution (N) (D)
of N for D..........DAN AND NAN CAN ADD DAN AND ANN

The effects of deletion, addition, and substitution on the "sense" of a phrase made up of three-letter words (codons) composed of only four kinds of letters (nucleotides) from the Roman alphabet. For purposes of illustration, alterations are made in the second position only of the first word, although in fact they could occur anywhere. As is evident, an addition or a deletion disrupts the subsequent sense of the message, which shows why the majority of "mutations" produce a reduction in the results of what is conveyed. Substitutions can produce a message that has some altered "sense" to it as is shown in the incestuous implications of the little family scenario depicted here.

GENETIC DRIFT

If the importance of the probable mutation effect is minor in comparison with that of natural selection, then the importance of genetic drift is practically negligible. Basically, genetic drift is the operation of chance as it influences the distribution of characteristics in a sequence of generations. This can easily be demonstrated by a simple example. Imagine a small, isolated band of Paleolithic hunters—say, six men and six women—encamped in a rock shelter at the edge of a game-filled valley. Hunting is good, and since they are the only people in that part of the world, their future seems well assured. A noticeable peculiarity among the men is that, whereas four of them have full heads of hair (as did their fathers, even in advancing age), the other two, having inherited a tendency to become prematurely bald, sport at best only thinning patches of what had once been long, healthy locks. On one unfortunate day, the roof of their rock shelter caved in, crushing most of the band of hunters—several of the women and all but two of the men. As luck would have it, both the unscathed men were the balding ones. On the strength of this characteristic alone, one could safely predict that the future males of the band would tend to be bald as well. Actually, this little example portrays a particular form of genetic drift called "the founder effect," but it nevertheless suffices to illustrate the logic of the general principle.

Evidently it was sheer chance which affected the future of the male contingent of this little group. Sheer chance is, in fact, a much more important element in determining proportions in a small group than in a large group. In a population of several thousand, for instance, the accidental loss of four men will not appreciably affect the relative proportions of characteristics transmitted to future generations, as it did in our example. Chance occurrences of this or any other sort cannot impart any consistent direction to evolution or account for an evolutionary trend. Since all major evolutionary developments are the results of forces which operate over very long periods of time, genetic drift is clearly of only local or incidental importance.

There are two reasons for mentioning genetic drift. First, it has been frequently invoked (too frequently, in fact) to account for changes that appear to have no adaptive significance. Second, throughout much of prehistory the characteristic human breeding population was just the sort of small, semi-isolated group in which genetic drift could have had its greatest effect. While it is difficult to point with certainty to human features that owe their existence to the operation of genetic drift, it would be equally foolish to deny that it had played any role in human evolution.

ORTHOGENESIS

Although orthogenesis is not really an evolutionary principle at all, it has been a prominent feature of former evolutionary schemes. We consider it here only so that we can convincingly renounce it. In the past, when evolutionary principles were dimly understood and did not seem adequate to account for the developments that the fossil record and contemporary organic diversity revealed, it seemed to some scholars that the visible evolutionary trends could be explained only by invoking mystical principles of unknown or supernatural guidance. Whether the forces invoked were cited as being "vital principles" or simply "evolutionary momentum," the result was the continuation of evolutionary development in a given direction for no obvious or logical reason. Straight-line development of this kind is what "orthogenesis" means.

There is no reason why development cannot occur in a consistent direction, provided that the selective forces which produce it continue to operate over long periods of time; clearly, this has happened in the evolutionary record. The main objections which the modern synthetic theory of evolution has to orthogenesis are its inherent implications that there is an unknowable force involved, and that development, once started in a particular direction, will continue of its own momentum even after the initial stimulus has ceased to operate. As is now realized, evolution is completely opportunistic. It is simply the accumulation of organic responses to continuing environmental stimuli. When these stimuli cease, the responses cease as well. In this sense, then, we must deny that orthogenesis is a valid evolutionary phenomenon.

PREADAPTATION

Preadaptation also is not really an evolutionary principle, but it *is* a useful way of looking at things, and it lets us understand some aspects of evolution that otherwise would not make sense. Organic form is not redesigned from scratch for each aspect of adaptive radiation. For example, the fins of a whale, the wings of a bat, the paws of a bear, the flippers of a seal, and the arms of a man are built upon exactly the same number and kinds of bones. Although these attain quite different end results, they are all achieved by modification of the same basic elements. It is obviously easier to redesign what is already there—even if there is some awkwardness and loss of functional efficiency—than to start over absolutely from scratch. The elements of the basic vertebrate limb, then, were already set in the limbs of the Paleozoic fish that served as the ancestor to the amphibians, which in turn gave rise to all of the reptiles and mammals that continue to exist today. That Paleozoic fish fin, then, was a preadaptation for the development of the arms and flippers and paws and wings of the ensuing adaptive radiation of the vertebrates.

Although preadaptation is the use of a structure that had originally been developed for one kind of use in a new situation, and for quite another kind of use, absolutely no foreknowledge is involved. Evolution is completely opportunistic, and it can never foretell that the future will bring circumstances in which a previously adapted condition will prove to be of use in another context. If, in fact, that actually proves to be the case, all well and good; but nothing ever takes place with that kind of *intent*. So when primates first evolved binocular stereoscopic vision, it was specifically to help them cope with the problem of moving through the forest canopy where depth perception was literally of vital significance, but there was no presumption that their remote descendants would ultimately use that capability to direct the path of motorized conveyances hurtling at high speeds over the freeways of Los Angeles.

That example actually is simply a situational transfer rather than the modification of a preexisting trait for use in a previously unforeseen set of circumstances, but it does serve to make the point. Our ancestral Ape Grade prehominid developed its quadrumanous clambering and arm-swinging suspensory locomotion for purposes of moving and feeding in the forest canopy. Since it is hard to hang horizontally, one of the consequences was the development of a vertical body plan, which has left its stamp on all the descendants of that creature. This was the preadaptation that determined that, when a terrestrial life way was subsequently resumed, the vertical body plan was retained. The result was the bipedalism that freed our hands from locomotor duties and allowed them to get involved in the use and manufacture of the implements that have been the key to our survival ever since.

NEOTENY

Some aspects of human form recall the infantile. In all newborn mammals, the head is relatively large and the face is relatively small in comparison with their final proportions when adulthood is reached. In the human adult, however, the head is still noticeably large and the face relatively underdeveloped, if the standards of comparison are such typical mammals as horses, dogs, rats, or sheep. An adult who displays the features of infancy is said to exhibit "neoteny," which literally means "the retention of the new."

The newborn human is remarkably helpless by most mammalian standards, and the period of juvenile dependency is noticeably prolonged. It is during this long, immature period of play and trial-and-error behavior that the growing human learns about the world. Human beings have carried a dependence on learned behavior to the maximum, and, conversely, almost everything that could be considered instinctive behavior has been eliminated. If this is a slower way of producing an adult who can cope with its world, it also results in an adult who can handle the new and different in much more effective fashion. In human beings, the teachability of youth is extended throughout the rest of life.

Some scholars have tried to view these various human capacities and features as simple manifestations of neoteny, a biological rendering of the poet Wordsworth's vision of the child as father of the man. The reliance on such imagery, however, encourages a tendency to confuse processes that are best treated separately. The adaptive advantage in the emphasis on braininess is obvious. The small human face—that is, small by prehuman or even Middle Pleistocene human standards—is not simply the product of retention of a supposedly advantageous juvenile condition, but rather the result of a failure to grow. Face development and brain development are under the control of completely separate sets of selective forces, and the proportion observed in comparing their relative development has no adaptive significance in itself. To invoke neoteny to account for that proportion, then, in fact explains nothing at all. When we look at the rest of the human body, it is obvious that the elongated legs and relatively short trunk of the human adult are actually the reverse of being neotenous. Applying the concept of neoteny to the human condition has tended to result in oversimplifications that have inhibited our ability to understand how modern human form has actually evolved.

SEXUAL DIMORPHISM

Sensitivity to the normal range of variation is important to the student of evolution, but sometimes the unstated referent can inhibit a proper appreciation of what that normal range might be. For example, when we try to project a reconstruction of the range of variation of a fossil population from a few scrappy specimens, we tend to do so with the model of the normal modern range in mind. To be sure, this is better than the approach that stresses the identification of an invariant typological ideal, but it does assume that the nature of human variation in the past was the same as it is in the present. That in itself is a kind of teleological assumption, and it has created more than a little misunderstanding.

Obviously, variation was just as normal prehistorically as it is now, but that does not mean that the *nature* of variation was the same. In fact, there is considerable evidence to suggest that it was not. This can be appreciated by a quick consideration of crucial differences in life way between Moderns and the people of remote antiquity.

In the postindustrial world, there are very few jobs that cannot be done equally well by males and females. Men are just as good at cooking and changing diapers as women are at driving bulldozers and laying bricks. Only in the physical acts of procreation and the mechanics of bearing offspring are there any evolutionary forces continuing to maintain the differences in male and female physiques.

This, however, was not true in the past. Before bottles and synthetic nipples were invented, a nursing mother was the sole source of nourishment for the infants upon whom the future of a population had to depend. In order to assure population continuity, the average human female spent her entire adult life either pregnant or lactating. Inevitably, the burdens of group defense fell upon the males. Also inevitably, the unencumbered mobility necessary for systematic, large-mammal hunting restricted such activities to males. The different requirements of male and female roles reinforced the maintenance of those secondary sexual distinctions in size, shape, and muscularity that are still visible, if no longer necessary for survival.

And if we project this kind of expectation back into the Pleistocene and beyond—to even simpler levels of technology—we should expect to find greater degrees of male-female distinction. At the very earliest stage of hominid development, when tools were rudimentary, the role of group defense must have led to the development of a degree of difference in size and muscularity between males and females that attained an order of magnitude not visible in the present or the recent past.

Among the terrestrial nonhuman primates—for example, baboons and gorillas—the selective pressure differences acting on the male involvement with group maintenance, as opposed to those acting on the female involvement with pregnancy and infant nurture, assured the development of a marked degree of male-female difference or sexual dimorphism. In both gorillas and baboons, males are literally double the size of females; they use that size and muscularity to ensure sexual access to as many females as they can control, and thereby to improve their chances at reproduction. Because they lack tools to help them in their efforts at group maintenance, selection has stressed the development of enlarged canine teeth—especially in the males.

Because the available evidence suggests that sexual dimorphism among the earliest hominids was maintained to an extent that was quite beyond what a study of the modern range of variation would lead us to expect, we should be alert to the possibility that they may have been using reproductive strategies more like those found in modern primates among whom sexual dimorphism is more marked than what we normally find among those modern human groups who managed to maintain a hunting and gathering mode of subsistence. But, unlike other terrestrial primates, the earliest hominids utilized tools in their various activities so that, although they may well have maintained a gorilloid degree of dimorphism in size and muscularity, they did not maintain a fully gorilloid degree of canine tooth-size dimorphism.

These considerations bring up a final matter that deserves some mention. If males and females differed to such a marked degree, then obviously any attempt to represent the group by a detailed description of only one specimen is going to constitute serious misrepresentation. In fact, it is just possible that some of the descriptions of different genera or species in the literature may be depictions of males and females of a single group. These points will all be kept in mind when we turn our attention to the actual fossil evidence of the stages of human evolution.

THE HUMAN ADAPTATION

Before embarking upon the final synthesis, wherein we apply the foregoing insights to the interpretation of the course of human evolution, we should give some attention to the age-old question, "What is man?" Because of recent complaints about the sexism embedded in our language, it is no longer even regarded as legitimate to phrase the old Biblical query in that form, but the sense of the question remains the same: What is this phenomenon called "humanity"? Definitions range from the realm of morality and philosophy to that of the pragmatic, functional, and physical. Depending on one's viewpoint, the more philosophical definitions can be characterized as displaying either dimensions of soaring insight or miasmas of vague verbalization. In any case, we leave this realm to other scholars with other purposes.

Attempts have been made to define humanity by means of specific, measurable, anatomical criteria, with the implication that such definitions are somehow "objective." As we have already seen, the recent revival of the approach of the medieval logicians has gained many adherents, although it continues to have the same kinds of problems that plagued its medieval exponents. In the eighteenth century, an eminent Dutch naturalist claimed that the human condition was indicated by the lack of an intermaxillary bone; within the last few years, anatomists have tried to claim that a brain capacity of at least 750 cc. is the minimum human criterion. The trouble with such criteria is that inevitable exceptions can be found, and that ultimately there is no necessary relation between them and the condition of being human. Definitions are created by people for their own convenience, and although they are essential for practical purposes, we should always keep in mind that, at bottom, they are arbitrary and subjective.

Viewed in evolutionary perspective, the most fruitful definition of humanity should be that which touches upon the most distinctive adaptation. Whereas arguments can be produced favoring the human brain in this regard, the social scientist can quickly counter this by pointing out that man does not survive by brain alone. However valuable it may be, brain does not serve as a substitute for experience. The most characteristic part of being human is the ability to profit from the accumulated and transmitted experience of other human beings. This can be regarded as the most important human adaptation; it is what anthropologists mean by the term "culture."

It is important to realize that culture, the primary human adaptive mechanism, is not a facet of human anatomy. In fact, some have referred to it as an "extrasomatic adaptation." Evidently, any attempt to define humanity on the basis of anatomy alone will be doomed to failure. This is not to say that human anatomy is unrelated to culture. As we shall see, quite the reverse is the case. But the anatomical correlates to the fact of cultural existence are reflections of an already-existing dimension, and they must be regarded as being after the fact rather than of primary importance in and of themselves. This is something we shall have to keep in mind when we look at the earliest hominids, where there is no preserved tangible evidence of culture, but where anatomical characteristics suggest that cultural attributes must have existed nonetheless.

This ability to transmit information and experience from one individual to another and from one generation to another is most clearly recognizable in languages and the records of their use. Unfortunately, however diagnostic language may be as an indicator of humanity, it leaves no trace in the archaeological or fossil record prior to the invention of writing. Since the vast majority of the happenings in human evolution occurred prior to this event, the prehistorian is put in the somewhat awkward position of trying to evaluate the humanity of our finds in the absence of the best criterion. Inevitably, the only tangible evidence for prehistoric human cultural capacity is prehistoric tools. This narrowing of the focus has even led some archaeologists to claim that the manufacture of tools should itself be the ultimate criterion of humanity, and to define people as tool-making animals. One consequence of this has been a heightened concern for the distinction between tool use and tool manufacture as a diagnostic criterion that is based on the feeling that a creature who merely selected suitably shaped natural objects as tools does not deserve to be called fully human. On the other hand, a creature who engaged in the regular modification of raw materials according to a set pattern could be appropriately elevated to the realm of the human. Contributing to this appraisal was the fact that prehistoric stone tools can be traced back to the point where the amount of shape modification is so rudimentary that they are little more than selected hunks of rock. Some archaeologists have felt that at this point we reach the boundary between the human and the prehuman.

There are several reasons why this latter assumption should be greeted with skepticism. First, field observation shows that modern chimpanzees engage in a little simple tool manufacture. Next, merely because the prehistoric creature in question was not shaping stone does not necessarily mean that it was not shaping perishable materials, which are manifestly easier to modify. Finally, the dichotomy between the simple selection of appropriately shaped raw materials and tool manufacture artificially creates categories out of what should be a zone of transition that is unrelated to the question of whether the survival of the user depended on the accumulated and transmitted experience of previous generations.

Properly speaking, the mere presence of shaped or unshaped tools in the archaeological record is of symbolic value. To the prehistorian, such a presence symbolizes the fact that the user was utilizing a dimension of patterned behavior of an order of complexity that is too great to have been discovered anew each generation. Such behavior could be acquired only by *enculturation:* the process of growing up in a social environment conditioned by the accumulated and transmitted learning of previous generations. Given no more than this, we must recognize that there almost certainly were rudimentary verbal clues in use which assisted in the enculturation process, but it is a much greater step to suggest that we are actually documenting the presence of even a simple form of what we would call language. Not too many years ago, it was fashionable to use the presence of tools as indicators of crude linguistic behavior, but the suspicion has recently grown that we were trying to read too much into the evidence. Early tools, in fact, remain remarkably similar through hundreds of thousands of years and extensive geographic stretches. Although language may not have been in the repertoire of their makers, we can certainly infer a creature who possessed the rudiments of culture in the anthropological sense and who could not have survived without it. Such a creature we can regard as a genuine hominid, but it might be stretching things a bit to call it a full human being.

chapter eight

Prehominids

True hominids did not suddenly emerge by a punctuation event from a remote prehistoric pongid. Instead, we can trace the picture of their obvious predecessors as it goes back to ever more primitive Primates, which in turn go back to the beginning of the Age of Mammals 63 million years (myrs.) ago, when they are visibly closer to a kind of generalized common mammalian stock than they are to the Primates of today.

THE GEOLOGICAL TIME SCALE

The major divisions of geological time, in which are preserved abundant evidence for prehistoric life, begin with the Paleozoic Era just under 600 million years ago. There are fossils in the preceding Pre-Cambrian, but so much distortion has occurred by the warping, folding, and erosion that has subsequently taken place that most depictions of the course of organic evolution begin with the Paleozoic Era. During this time we can see the emergence of primitive and then more evolved fishes. Some of these develop pelvic and pectoral fins with the same number and kind of bones in them as our own arms and legs. From these fish, full-fledged amphibians emerged. By the end of the Paleozoic 240 myrs. ago, they had proliferated and diversified in the extensive swampy areas that produced the coal deposits that are of such importance today. Among those amphibians is one that is transitional in many respects to the reptiles.

The Mesozoic Era is the second of the large geological divisions that contain the fossil evidence for organic evolution; it has informally been called the Age of Reptiles, a time when land-going adaptations were perfected. Whereas the moist and permeable skin of the amphibians permits the evaporation of body fluids, which can lead to fatal dessication if individuals are removed from moisture for prolonged periods, the scaly skin that was acquired by the early reptiles allowed them to pursue a life that was no longer dependent on proximity to the ponds, swamps, and

HOLOCENE	Recent	10,000 years
CENOZOIC	Pleistocene	2 million
	Pliocene	5 million
	Miocene	24 million
	Oligocene	38 million
	Eocene	55 million
	Paleocene	63 million
MESOZOIC		240 million
PALEOZOIC		590 million

The divisions of geological time.

streams which were, and still are, the milieu of the amphibians. The other critical reptilian adaptation was the development of eggs with shells. Because these could be laid on land, this development marks the final separation from that dependence on the presence of standing water which characterizes the amphibian adaptation.

Within the Mesozoic we see the proliferation of that particular group of reptiles that has so captured the imagination of the public: the dinosaurs. The reptiles, in fact, went through a major adaptive radiation. Some forms returned to the sea and redeveloped adaptations for swimming; others developed the power of flight; many differentiated into various herbivorous and carnivorous terrestrial forms. One small member of the dinosaur group underwent an adaptation whereby its scales became transformed into feathers. Indeed it was intermediate in many respects between the class **Reptilia** and the class **Aves**, or true birds. This particular form, *Archaeopteryx,* clearly was ancestral to the birds. In the Mesozoic, also, the therapsid or mammal-like reptiles provided the base from which true mammals emerged to become the dominant form in the succeeding Cenozoic Era, which, appropriately, is known as the Age of Mammals.

REPTILIAN EXTINCTIONS

What caused the extinction of the ruling reptiles is still a matter of much vigorous debate. Many forms of both marine and terrestrial life disappear at the Cretaceous/Tertiary boundary, which marks the transition from the Mesozoic to the Cenozoic 63 million years ago. Some have postulated that a comet or asteroid collided with the earth, and that the ensuing dust cloud blocked out the rays of the sun long enough to cause the death of quantities of plants. According to this scenario, animal forms from microorganisms to dinosaurs, dependent on those plants, died off in turn.

Although it cannot be denied that such a sudden and dramatic catastrophe *could* have occurred, many geologists and paleontologists are skeptical and try to use a combination of more expectable means to account for the many extinctions that clearly took place late in the Cretaceous—the last period of the Mesozoic Era. Changes in the positions of the major continents may have altered previously prevailing weather patterns. Increased volcanic activity accompanying that continental repositioning may also have had its effect. In any case, whether the cause was extraterrestrial or the result of a configuration of less catastrophic geological phenomena, the result undoubtedly was the disappearance of many members of the previously ruling reptiles and the opening of a series of niches for exploitation by the emerging class of mammals.

PRIMATES

Among these were representatives of the order Primates. As mammals, these were characterized by the possession of fur for insulation, warm blood as an indication of a constant elevated metabolic rate, and the production of live offspring which are nourished by milk from mammary glands. They are not differentiated from the rest of the mammals by any one absolute thing, but rather by the configuration created by the combination of a series of retained primitive and emergent derived features. Whereas in many mammals the number of fingers and toes is variously reduced, in the Primates all five are retained on both the fore and hind limbs.

The grasping capabilities that this reflects, which are of use in manipulating food items and also in climbing activities, serve to point up what it was that shaped the Primates as an order distinct from the rest of the Mammalia—that is, a tree-going or arboreal life way.

As the early Cenozoic mammals underwent their adaptive radiation, some became terrestrial herbivores, some became carnivores, one branch developed the capabilities of flight and gave rise to the bats of today, others went back to the sea as whales and their relatives, and the Primates exploited the arboreal ecological niche. All of those retentions and developments reflect adjustments—adaptations—to the problems that face an arboreal animal, and the stamp of that arboreal heritage is visible in all of the Primates alive today, including ourselves.

An arboreal environment puts different stresses on its inhabitants than does a terrestrial habitat, and the Primates all show a series of consequent adaptations. There is a reduction in the reliance on the sense of smell, and the snout tends to be shortened as a result. Locating major supports and hand holds at a distance tends to be done visually. The importance of binocular stereoscopic vision for depth perception means that the eyes tend to be moved forward toward the front of the skull where their fields of vision overlap, rather than being laterally placed, as in most animals.

Finally, movement through a forest canopy requires making constant decisions. On the ground, even something as relatively witless as a rhinoceros can plod ahead, step after step, secure in the unreasoned faith that solid ground will continue to be there to support it. In a tree, however, the situation has to be constantly assessed and choices have to be made. Inevitably, the mechanism for making the choices—namely, the brain—tends to be enlarged beyond what is the case in nonarboreal creatures of comparable size. From early in their history, Primates have tended to be the brainiest of animals, even though a Paleocene Primate would look rather underendowed in that respect compared with practically all comparably sized mammals today. But in the context of the Paleocene, the creatures we identify by these comparative means as Primates clearly have a cerebral edge over the other representatives of the animal world.

In the later stages of Primate evolution, the hominid clade capitalized upon this cerebral expansion ultimately using it as the device for creating a unique kind of adaptation—the cultural ecological niche—which, in turn, engendered selective pressures that led to further adaptive increases in brain size. This business of being able to use a trait developed for one set of circumstances to take advantage of a completely different situation is a classic example of preadaptation. In this case, the presence of both an expanded brain and a developed grasping and manipulative capacity combined to produce the cultural milieu that has become the human ecological niche.

PALEOCENE PRIMATES

In the Paleocene, which began 63 million years ago, there are a number of primitive kinds of Primates, although there is some question as to which, if any, of the known forms gave rise to the Primates of later periods. Fossils of a rat-sized Primate, found in the Rocky Mountain area of the United States in the Early Paleocene and possibly the Late Cretaceous, have been identified as belonging to the genus *Purgatorius*. Members of this group are both early enough and of general enough form to give us an idea of what the Primate common ancestor must have looked like, whether or not *Purgatorius* actually is that common ancestor.

Other Paleocene Primates, exemplified by the form *Plesiadapis,* show traits which exhibit a convergence with the rodents, and it is probably no accident that they become extinct when the rodents subsequently undergo their own adaptive radiation. The rodents, as we know, are not only highly successful as gnawing, seed-eating creatures, but they are also capable of reproducing themselves in quantities and at a speed with which no Primate can compete.

The Primates of the Paleocene and the succeeding Eocene all qualify as belonging to the Prosimian Grade. Whichever particular form actually proves to be ancestral to the Primates of more recent times, it is clear that it was some sort of Prosimian that gave rise to the later monkeys, apes, and hominids.

EOCENE PRIMATES

During the Eocene, which started 55 million years ago, not only do we find Primates that serve as good candidates for the ancestors of more modern Primates, but we can also see them change through time and take

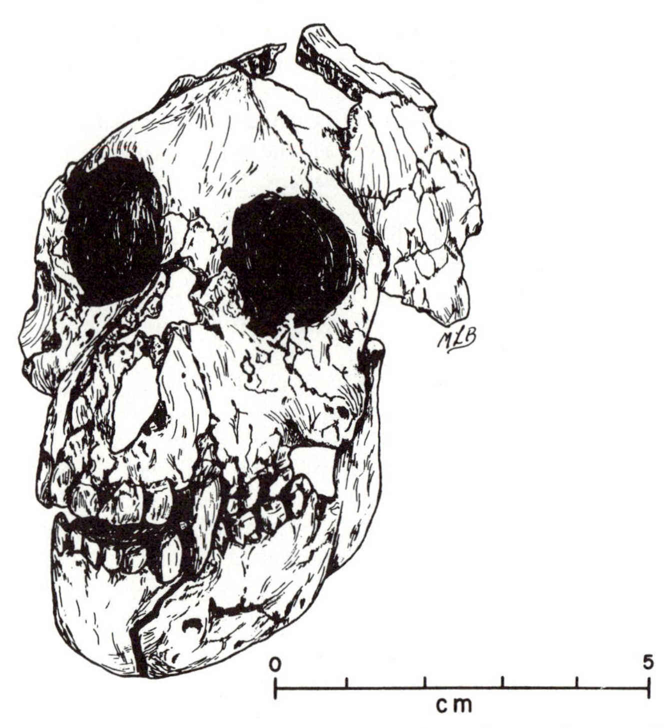

Aegyptopithecus zeuxis. (Drawn by M. L. Brace from a photograph furnished through the courtesy of Professor E. L. Simons of Duke University.)

on a slightly more modern aspect as they do so. For the past fifteen years and more, Philip D. Gingerich of the University of Michigan has been leading annual expeditions to the Bighorn Basin of Wyoming. There, among the thousands of Eocene fossils he and his coworkers have uncovered, Gingerich has found some hundreds of specimens of a Primate of the genus *Cantius.* This was a small, rat-sized creature that was comparable in overall form and general level of adaptation to a modern lemur—that is, somewhat less than a full-scale monkey but more than a non-Primate such as a tree shrew.

The Bighorn section is rich and detailed, and there are specimens representing *Cantius* from at least 20,000 year intervals in an unbroken sequence for several million years. This in itself is one of the more remarkable fossil demonstrations of mammalian continuity. But what makes these finds particularly interesting is that the *Cantius* specimens change so much from the bottom of the sequence on upwards that, by the top, they have to be regarded as an entirely different genus, *Notharctus.*

Notharctus fossils have been known for more than a century, and, throughout that time, they have been recognized to represent a candidate for just the sort of lemur-like creature from which monkeys, apes, and, ultimately, humans are derived. Whether or not *Notharctus* is that specific ancestral form, its emergence from the abundantly documented *Cantius* predecessors has given us a splendid picture of early Primate evolution from its most primitive and almost insectivore-like origin up to the threshold of the Monkey Grade.

OLIGOCENE PRIMATES

Although the Eocene Primates, especially the early ones that represent the most likely candidates for the ancestors of the Primates living today, are most abundantly represented in North American fossiliferous deposits, the Oligocene continuity is best demonstrated by the fossil record in Africa. The work done over the last twenty-five years by the American paleontologist Elwyn L. Simons in the Fayum region of Egypt west of the Nile has provided a rich assortment of Oligocene fossils. The Oligocene extends from 38 to 24 million years ago. The most justly famous and important of these discoveries, dating from between 33 and 34 million years ago, is *Aegyptopithecus zeuxis.*

Aegyptopithecus is a relatively heavy-bodied arboreal quadruped, about the size of a fox, and it is clearly at what could be called the Monkey Grade of organization. Brain size has increased, relatively speaking, over what can be seen in *Notharctus,* and the eye sockets have rotated more toward the front of the skull, which suggests the development of overlapping stereoscopic vision.

Details in the crown patterns of the molar teeth clearly foretell the pattern that is later evident in the apes of the Miocene and of today, as well as the whole hominid line. But the presence of a well-developed tail and the clearly quadrupedal adaptation indicated by the form of the arms and legs keeps *Aegyptopithecus* firmly anchored in the Monkey Grade. All told, *Aegyptopithecus* is a splendid representative of just what we would expect the Monkey Grade ancestor of subsequent apes and human beings to have looked like.

MIOCENE PRIMATES

The Miocene began 24 million years ago, and within the next 6 or 7 million years, the continent of Africa moved north far enough so that the mammals that had been evolving there could expand their ranges into the rest of what we now regard as a single entity: the Old World. As the Miocene proceeded, ape-like Primates of evident African origin came to occupy a range all the way from Europe to China. Today their scattered remains can be found in the regions between wherever there are fossil-bearing deposits of the right age.

The best candidate for the source of this group of Miocene Primates is *Proconsul africanus,* another of the pivotal discoveries made by Mary and Louis Leakey. The Leakeys found this specimen in 1948 on Rusinga Island, located in Lake Victoria at the western edge of Kenya, in deposits that can be dated to 18 million years. Details of skull, face, jaw, and tooth form make this an ideal candidate to represent the ancestor of the subsequent apes, but the form of the arms, legs, and body shows that *Proconsul* was only at the threshold of, but not yet fully in, the Ape Grade of quadrumanous, arm-swinging clambering.

During the Middle Miocene, the widespread ape-like forms all bear a noticeable resemblance to one of the very first prehistoric Primates to have been discovered: the 1856 French find, *Dryopithecus fontani.* This allows us to refer to the bulk of the Miocene apes as "Dryopithecines," a term that can be used to include the original *Dryopithecus* from France, *Graecopithecus* from Greece, *Ouranopithecus* from Hungary, *Ankarapithecus* from Turkey, *Sivapithecus* from Pakistan and India, a similar genus in China, and, arguably, even the original *Proconsul* as well as *Kenyapithecus* in Africa.

The *Sivapithecus* discoveries from the Late Miocene (8 million years) of Pakistan and related forms found in southern China are particularly interesting because the details of the anatomy of their facial skeleton and the way their teeth are implanted show obvious similarities to the recent Southeast Asian orangs. Enough of the postcranial skeleton of *Sivapithecus* has been found to indicate that the elongation of the arms associated with the quadrumanous clambering mode of locomotion had already been achieved. *Sivapithecus* was fully in the Ape Grade, and is the obvious candidate as the ancestor of the modern orangutan of Sumatra and Borneo.

In Africa, the Dryopithecine descendants of *Proconsul* in the Middle Miocene show clear signs of being ancestral to both the modern African apes and the first true hominids—the Australopithecines—which means they can be considered ancestral to us. In the succeeding geological period, the Pliocene—which began 5 million years ago—there is a dearth of

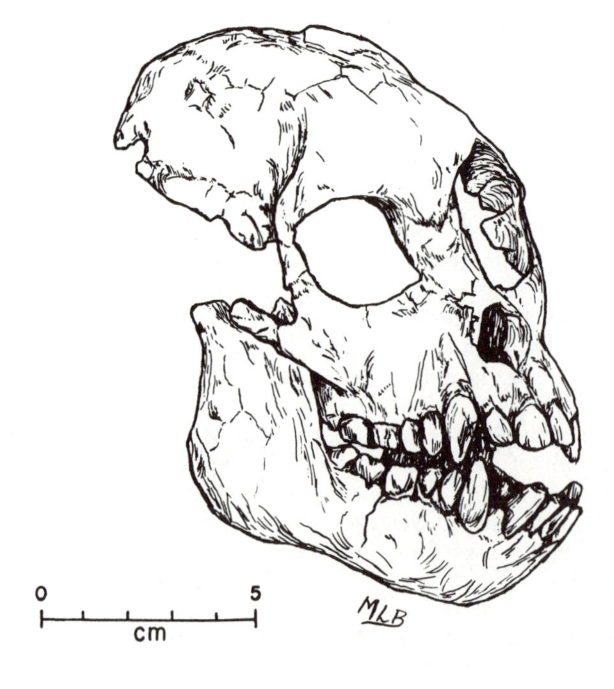

Proconsul africanus.
(Drawn from a cast by M. L. Brace.)

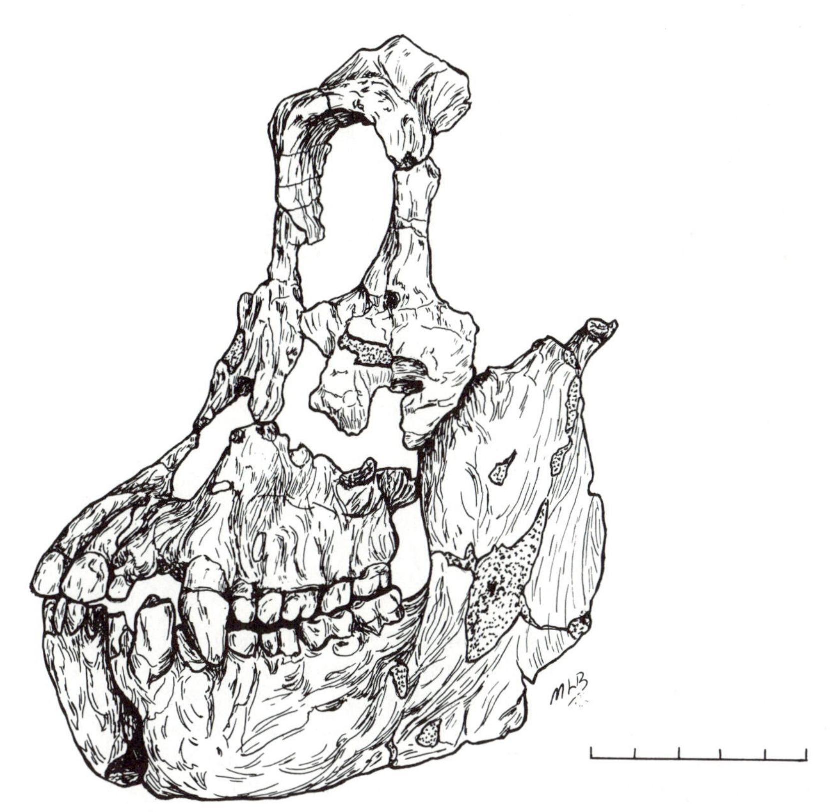

Sivapithecus indicus. An 8-million-year-old specimen from Pakistan, GSP 15000. (Drawn by M. L. Brace through the courtesy of Professor D. R. Pilbeam of Harvard University.)

pongid fossil material, but by this time we do have the first full-scale hominids—the Australopithecines—and the continuity from that time on is relatively unbroken. This provides the subject matter for the remaining chapters of this book.

chapter nine

Culture as an Ecological Niche

THE PLEISTOCENE

With the presence of stone tools considered as sufficient to establish the existence of a culture-dependent creature—a genuine hominid—we must then recognize the fact that some such form must have been in existence in the Early Pleistocene, some 2 million years ago. The Pleistocene is the geological period that contained the recently ended ice age. Since this was the period during which all of the major events of human evolution took place, it has necessarily been the focus of considerable attention on the part of the anthropological world.

Recently, geophysicists have utilized a variety of ingenious techniques to establish the age of strata in the recent past. Recognizing that various radioactive elements "decay" into stable end products at fixed rates, they have measured, in material taken from crucial geological layers, the ratio of certain of these radioactive elements to their stable end products. Since the ratio discovered is proportional to the length of time during which the process has been going on, it serves as a measure of the time elapsed since the layer in question was formed. By these means, the Pleistocene has been calculated to extend from about 2 million down to some 10,000 years ago.

Perhaps this makes Pleistocene dating seem simpler than it is. Actually, there are a great many knotty problems connected with it. For instance, finding suitable mineral specimens to use for this kind of analysis is beset with difficulties. Indeed, only a very limited number of layers have been pinned down in this way. The rest have been tentatively fitted in by extrapolations based on the fossil animals contained. As a result, a great deal of uncertainty still remains concerning such problems as the correlations and relative ages of layers in South Africa compared with those in Indonesia, or Europe with China. Despite all these uncertainties and inaccuracies, however, the temporal dimensions of the period during which humanity evolved are beginning to emerge.

During the Pleistocene there were periodic major onsets of glacial conditions in the northern hemisphere. Geologists used to regard the whole period as being taken up with glaciation, and they thought that there were only four clear-cut Pleistocene glaciations. But it has recently become apparent that episodes of glaciation have been recurring at 100,000-year intervals for somewhat less than the last 1 million years, and that there may have been as many as a dozen glaciations over the past

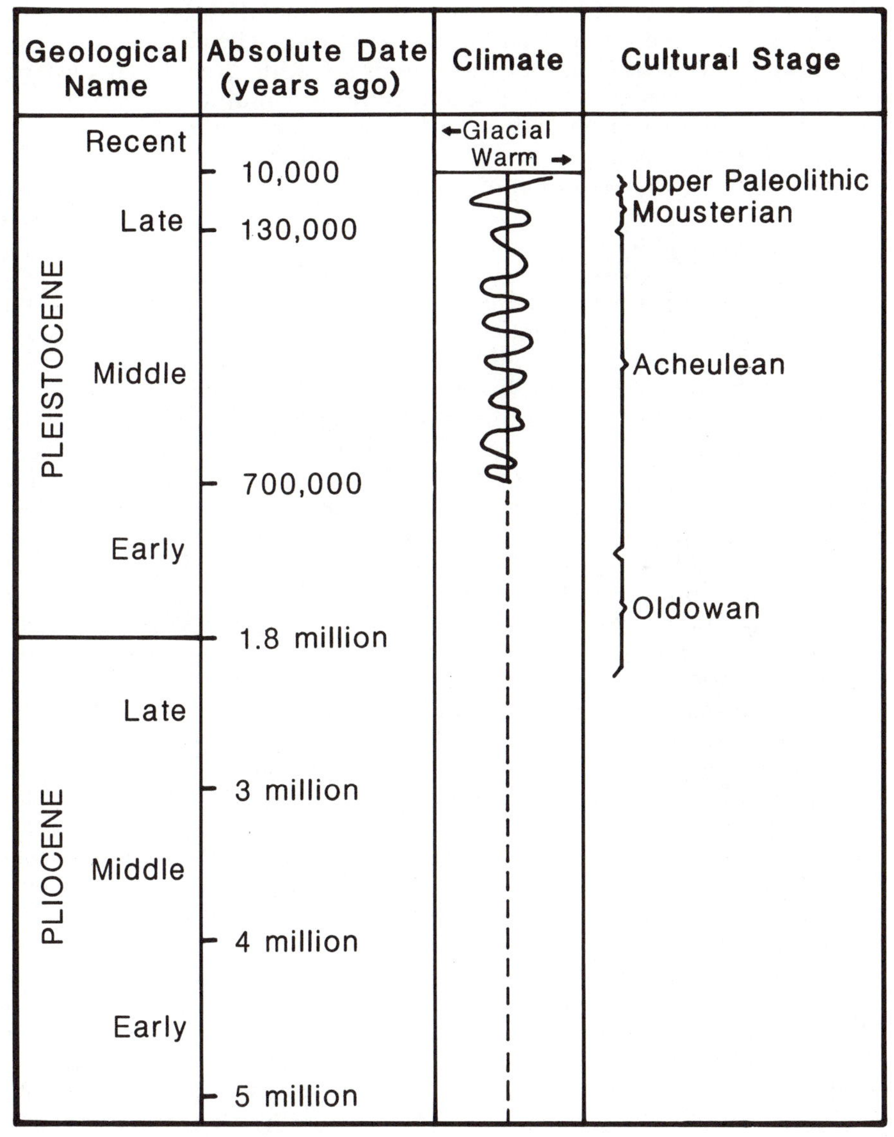

The divisions of the Pleistocene.

million plus years. The earlier part of the Pleistocene was characterized by milder climatic conditions, although there were indeed fluctuations between drier and better-watered periods of time in the areas inhabited by the early hominids. The accompanying chart displays the divisions of the Pleistocene, their approximate duration, and the associated cultural stages displayed in the archaeological record.

The most consistent and reliable evidence for hominid existence throughout this time is seen in the form of chipped stone artifacts. The tool-making traditions practiced during the Pleistocene have been given the label Paleolithic (old stone), and reveal the fact that the characteristic modes of subsistence were hunting and gathering. The change to a food-producing way of life did not begin to occur until after the Pleistocene was over. This was one of the most significant cultural events that ever occurred, and it has profoundly changed the nature of the selective forces acting on human populations during the last several thousand years. We shall return to a consideration of these matters in a later chapter.

PALEOLITHIC TOOLS

The Paleolithic is further divisible into Upper, Middle (tentatively), and Lower segments of very unequal length; 90 to 95 percent of the record is included in the Lower Paleolithic, during which cultural change was extremely slow and cultural diversity apparently at a minimum. We say "apparently" because all we know of the Lower Paleolithic is the few stone tools remaining and the bones of the animals that people ate. While these indicate that the gross dimensions of human life were much the same from one end of the inhabited world to the other, it is perfectly possible, and even likely, that minor cultural differences of an unknown nature flourished in different areas. Stone tools form only a small component of the total cultural repertoire, which, we should remember, includes far greater quantities of perishable items and, even more important, dimensions of verbalization, knowledge, and social behavior that leave no record. Potential diversity of this sort notwithstanding, however, it is still possible to state that major and basic facets of human adaptation were substantially the same from one end of the Old World to the other. Whether in South Africa, Europe, Asia, or Indonesia, game was hunted by the same techniques and processed by chipped stone tools which were virtually identical over vast areas.

The increasing local diversity visible in the stone tools of the Middle and Upper Paleolithic will be treated later, but at the moment we are concerned with the Lower Paleolithic, and specifically with its earliest phases. According to current indications, the oldest cultural remains in the world, and consequently the earliest nonanatomical evidence for human existence, come from East and South Africa. The tools on which this judgment is based are of the crudest recognizable sort, and, were it not for the location in which they have been found, it would be almost impossible to prove that they were indeed the products of deliberate manufacture. But, occurring as they do by ancient lake margins and out

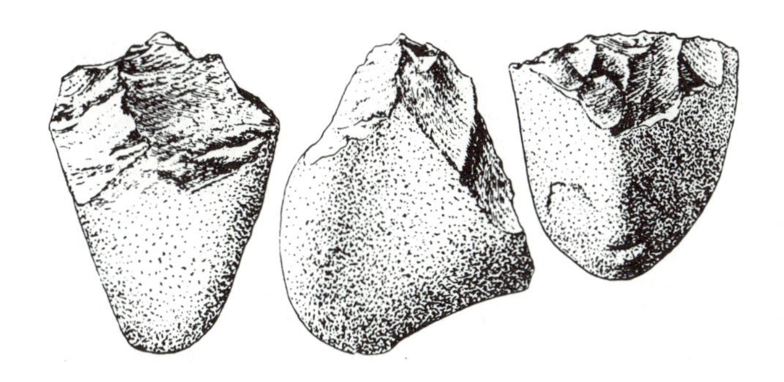

Pebble tools of the Oldowan type from Bed I of Olduvai Gorge, Tanzania. (By courtesy of the Trustees of the British Museum of Natural History.)

on the plains miles from the nearest potential rock outcrop or natural source, we can only conclude that they were deliberately transported there. The discovery of such objects in a rocky stream bed or among the countless thousands of pebbles or cobbles which natural water action has battered and fractured is something that could take place by perfectly natural means. But found in a fine sedimentary deposit amidst the dismembered remains of extinct animals, they clearly indicate the activities of an ancient hunter.

In their crudest form, these early tools resemble river pebbles from which a flake or two has been knocked off, creating an edge or point. This is the source of the type designation by which they are known: "pebble tools." Because of the simple nature of the flaking, it is impossible to say whether it was done deliberately, or whether it had occurred naturally and the ancient hominid had simply chosen the stone from among others for this reason.

We mentioned controversy over the importance of tool manufacture as distinguished from tool use in a previous chapter. Although this controversy will probably never be resolved, the important thing to remember is that the presence of these tools, whether chosen or fashioned, signifies the presence of a creature whose life depended upon the behavioral complex of which these tools were a part—that is, whose survival depended upon the presence of culture.

There is another crucial fact to be recognized, and this is that there is evidence for no more than one basic cultural tradition in the Early Pleistocene. By tracing this up through time, we can see that this tradition is the direct ancestor of all succeeding cultural traditions, which makes it, among other things, the remote parent of this book. A single original cultural tradition strongly suggests that only one organism developed culture as a necessary condition for its survival.

ECOLOGY AND CULTURE

Ecology is the study of the life ways of species; the total life way of a given species can be called its ecological niche. In the human case, all facets of living are conditioned by culture, and it is therefore justifiable to regard humans as inhabiting a cultural ecological niche. Another ramification of evolutionary theory, which has been called the competitive exclusion principle, states that no two organisms can continue to occupy the same ecological niche. In a sense, this is stating the obvious, although, as in so much else, it is not always obvious until stated. The point is this: Culture is a single ecological niche within which one would expect to find only a single species. Because there is only one cultural tradition visible in the Early Pleistocene, the probability is greatly increased that there has been only one hominid species since the beginning of that cultural record, and that the hominids at different time levels are lineally related.

But if culturally dependent creatures of different time levels are probably related, what about the occurrence of two different forms of hominid at the same time? How could that occur when competition within a niche—the cultural ecological niche, in this case—tends to eliminate one of the competing forms? We can only guess that at its earliest stages of development, culture was less effective and was consequently a less-than-all-pervasive determinant of the nature of the total life way that it eventually came to dominate. And even though the skeletal evidence from the very earliest hominids goes back more than a million years before the first stone tools, that skeletal evidence suggests that those hominids were dependent upon some aspects of culture—presumably of a perishable nature—for their survival. From the Late Pliocene and Early Pleistocene skeletal remains, it would appear that more than one kind of hominid made an effort to enter the cultural ecological niche, but, in the long run, only one continued.

The term cultural ecological niche, then, is an abstraction that we can use for our own convenience. But, as biologists have found when they have tried to define other kinds of ecological niches, we must be careful lest we raise a convenient abstraction to the level of a typological essence. The concept of ecological niche is a convenient way of generalizing about organic adaptation—and no more. We should not allow it to take on an idealized existence of its own. For making broad generalizations, such categorizations can be useful, but we must continue to be aware of the particular and separate factors that can exert very specific influences. If the general capacities of modern *Homo sapiens* have been conditioned by long-term adaptation to the cultural ecological niche, some obvious differences—skin color, nose form—are the product of regional differences in the intensity of selective forces that themselves are distributed without regard to human cultural capacities. In short, the cultural ecological niche is a useful concept, but we have to be careful that it does not turn around and use us.

Just as the evidence for cultural development can be viewed in terms of sequential stages, so can the picture of human physical development be treated in similar fashion, although its stages are considerably more tentative because of the fragmentary nature of the record and the long gaps between crucial specimens. For the earlier stages, the fossil evidence is spotty and widely separated, and the relative ages of the various localities are bitterly contested. For most of the later stages, the evidence is more complete, the dating more reliable, and the association with the cultural record more certain. In spite of this improvement, however, there are still many problems and disputed pieces of evidence, and many professional scholars are vehemently opposed to the implications of the solution about to be offered. Speculation is rife, and since no interpretation can be more than that, the best justification for the interpretation that follows is that it is consistent with both cultural and evolutionary theory.

chapter ten

The Australopithecine Stage

HOMINID CHARACTERISTICS

Africa, which has provided us with the earliest archaeological evidence for human existence, has also yielded the earliest hominid skeletal remains. The term "hominid" is a colloquial version of the technical term **Hominidae**, the taxonomic family to which human beings belong. The closest human relatives within the order **Primates** are the living anthropoid apes. These belong in the family **Pongidae,** which, together with the **Hominidae,** is included within the superfamily **Hominoidea**. To simplify reference, the term "pongid" is generally used to designate anything that is more ape-like than man-like, while the term "hominid" is generally used to mean "taxonomically included within the family which harbors human beings proper."

The initial specimen, *Australopithecus africanus,* has given its name to a whole group of what are variously called "ape-men," "man-apes," "near-men," or even "primitive men." They include the Transvaal finds by Dart, Broom, Robinson, and others. The category can also be extended to include "Zinj" and other specimens from Bed I at Olduvai, many of Richard Leakey's discoveries east of Lake Turkana, material from the Omo River valley in southern Ethiopia, and the Pliocene finds of more than 3 million years antiquity made by Johanson in the Afar Depression of central Ethiopia, and by Mary Leakey, at Laetoli, twenty-five miles south of Olduvai in Kenya. Broom and others have attempted to elevate these to subfamilial status within family **Hominidae** and to call them **"Australopithecinae."** This would allow the retention of all the separate generic names as valid taxonomic units, but, in terms of the present analysis, this course of action seems untenable. It is still useful to refer to the group as Australopithecines, and, as we shall see, they can properly be regarded as the first stage in human evolution.

The differences of opinion concerning Australopithecine taxonomy revolve around the criteria that are considered important for purposes of classification. For Broom and Robinson, visually perceived differences and similarities are of prime importance, and taxonomic status depends on how many there are of each. Thus, the Australopithecines are ape-like in the possession of small brain cases, big molar teeth, facial projection, and a number of other features, but they differ from the apes in that they lack projecting canines, have a downward instead of a backward facing *foramen magnum* (hole at the base of the skull where the spinal chord enters), and have a shortened and expanded ilium (hip bone) and other related characters. In this latter regard, they resemble humans more than apes. Broom and Robinson (and others) consider that this balance of human and pongid features justifies a taxonomic position distinct from the one occupied by modern humanity, and of greater importance than merely either a specific or generic distinction. It is a position that has to be given thoughtful consideration, even if we do not accept it at the formal level. Note, however, that the Australopithecines can still be formally considered within the Family **Hominidae**, and hence they are called hominids.

Without denying that the balance of Australopithecine characteristics lands somewhere between the pongid and hominid categories, it is worth pointing out that not all characteristics are of equal importance to the survival of the organism. It would seem that those characteristics which

Artist's conception of an Australopithecine. Note the simple look on his face and the presence of fur. This is sheer fancy, since there is no way of telling facial expression or other aspects of the soft parts of the body from the fossil record by any direct means.

have the greatest adaptive significance should be the ones to determine the major taxonomic category. In the case of the Australopithecines, two characteristics outweigh all others, but it is not so much the characteristics themselves that are of import as what it is they signify. The first is the nonprojecting canine, and the second is that the Australopithecines were erect walking bipeds. Although these features have been recognized for years—in fact, they have largely contributed to the inclusion of Australopithecines within the category hominid—their full significance has often been misperceived, and arguments about interpretation continue among the experts.

CANINE TEETH

To take up the first of these characteristics, it is most suggestive that the Australopithecines and subsequent human beings, alone of all the terrestrial primates, do not have projecting canine teeth. Typically, terrestrial primates have greatly enlarged canine teeth (witness the baboons), since, as small, relatively slow creatures, they could not survive the depradations of a variety of carnivores without some effective means of defense. At the same times, it has recently become apparent that the extraordinary enlargement of *male* canine teeth in many primates is due more to competition between males for access to females than it is a response to the threat of predation. Even so, female baboon canines project well beyond the level of the rest of the teeth in the dental arch, and it can be painful indeed to be bitten by a female baboon. In human beings, however, canine teeth do not project beyond the level of the other teeth and consequently play no role either in defense against predation or in competition with other males in gaining access to available females. In the human case, the lack of canine projection is clearly compensated for by the use of hand-held tools employed for a variety of purposes, including

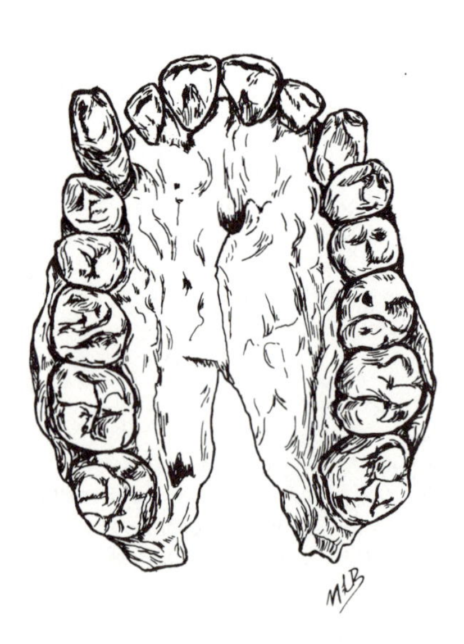

The palate AL 200 from the Hadar site in the Afar Depression in north central Ethiopia, approximately 3 myrs. ago. All of the teeth are large and robust, and there is an ape-like gap or *diastema* between the lateral incisor and the canine. The canine, while large, does not project in a functional fashion above the occlusal level of the rest of the teeth. The apparent projection of the right upper canine is because it had partially extruded from its socket prior to fossilization. The normal condition is shown by the canine on the other side. Even though the canines do not project, the form of this palate and these teeth is almost more pongid than hominid. All told, it is sufficiently intermediate that a decision as to which it should be called cannot be made on this part of the skeleton alone.

defense and aggression, and it is difficult to interpret the Australopithecines in any other light.

BIPEDALISM

The other point to consider is that the Australopithecines were bipeds, as can be seen from the anatomy of the pelvis and leg and the placement of the skull. To be sure some recent studies have shown that a number of features visible in the finger and toe bones, as well as in the arms and shoulders, suggest that the Australopithecines retained climbing capabilities of a far more specific nature than is true for the odd tree-climbing modern human. Because of this, it has been suggested that they may well have continued to use trees as places of refuge. Although this is perfectly plausible and even likely, it does not change the fact that their lower appendages, even while retaining a number of pongid and transitional aspects, were those of a terrestrial biped whose style of locomotion on the ground was far more like ours than it was like that of chimpanzees or gorillas.

Even though they were indeed bipeds, this has sometimes been taken to imply more than could be justified by the evidence. It was fashionable a generation ago to envision the Paleolithic hunter bounding across the grassland on his long, straight legs, as though bipedalism were somehow the most efficient, speediest, and "best" possible way of getting around. There are even some artists' depictions of Australopithecines sprinting across the savannas in hot and threatening pursuit of herds of antelope. To any who may still harbor the lingering residue of such an illusion, I suggest that you seriously consider the vision summoned by an irate adult *Homo sapiens* in hot pursuit of a thoroughly frightened *Felis domesticus* (house cat). As anyone who has ever engaged in this exercise in futility knows, the point is obvious: as a mechanism for high-speed locomotor efficiency, hominid bipedalism is ludicrous. Obviously, a creature who cannot even catch a small cat has proportionately less chance of getting away from a large one, and it is certain that such felines as leopards and lions, as well as a variety of pack-hunting canines, must have posed a constant threat to the survival of any savanna-dwelling Primate during the Plio-Pleistocene.

The only possible excuse for the development of hominid bipedalism is that it allowed for the development of compensating features (but clearly not formidable canines). Because the main functional correlate of bipedalism is that the hands are freed from any involvement in the locomotor process, it would seem that the hands must have been used in wielding a nonanatomical defensive mechanism. Given a creature lacking in dental defenses and pathetically slow of foot, we could postulate the existence of culture even *without* any direct evidence. Culture, in the form of recognizable but rudimentary stone tools, does exist right back to the beginning of the Pleistocene 2 million years ago, and it is evidently ancestral to the subsequent traditions of the Middle and Upper Pleistocene, where it is clearly the manufactured product of our own forebears. Furthermore, the hominids at that time were all recognizable as Australopithecines.

Not very many years ago, we could feel a sense of gratification in the association of the earliest demonstrable evidence for culture with hominids whose anatomy indicates that they could not have survived without it. As we have since discovered, however, things are not quite so simple. On the one hand, in the Hadar region of the Afar Depression in north-central Ethiopia and at Laetoli, just south of Olduvai Gorge in Tanzania, there are abundant remains of recently discovered Australopithecines that date back more than 3 million years. This is well back into the Pliocene and about a million years before the first stone tools. And shortly after the first tools did put in their appearance at the onset of the Pleistocene, there is clear evidence for two distinct kinds of Australopithecines. Finally, by 1.3 million years ago, we find a more developed stone tool assemblage, Australopithecines, and full-fledged Pithecanthropines. Given these data, we can legitimately ask, "What does it all mean?" Can we interpret it in such a way that we can get some sense of what was going on? In the current vernacular, what "scenario" does this suggest? Admittedly, many pieces of information are not mentioned in the preceding sketch that might bear on the question, but many of them are so incomplete that they are just as likely to confuse as to clarify. Despite that, it is indeed possible to sketch the outlines of a synthesis.

THE ORIGIN OF HUNTING

One of the keys to our understanding is the realization of what the evidence does and what it does not tell us concerning the dimensions of the cultural ecological niche. As we have noted, the presence of manufactured tools is direct evidence for the existence of culture, and this in turn has led us to anticipate the presence of those other aspects we have come to associate with it: expanded dependence on learning, symbolic behavior, planning for the future, and the like. These are inferences, however, and the mere presence of tools per se provides no more evidence in their support than does the existence of a biped that does not have projecting canine teeth. Indeed, many have come to feel that we have invested the Australopithecines with more of the distinguishing attributes of humanity than the evidence really warrants.

On the other hand, the context and associations of the tools can tell us more than the not-so-insignificant fact that their makers depended on culture for survival. The shape of the worked stone pieces themselves is not particularly instructive. The early ones are about the size of a dinner roll, with a flake or two knocked off to make a pointed or chisel-like business edge. And although no one doubts that the object in making the tool had been to produce a working edge that could be employed for some important function, there has been much less certainty about what that function actually was. Guesses concerning their use have ranged from suggestions that they were employed for digging roots, for shaping wooden objects, or for serving as hand-held weapons. Artists' reconstructions variously depict Australopithecines facing marauding carnivores or even each other with pebble tools clutched in hand in dagger-like fashion, or sprinting across the African savanna in pursuit of an antelope or gazelle while wielding their tools in a threatening manner.

One of the problems in dealing with the events of prehistory is that we can never go back and put our interpretations to any immediate test. Instead, we have to rely on circumstantial evidence. Our conclusions, then, are tentative at best, and we can only offer them with varying degrees of probability instead of proof. With perserverance, awareness, and luck, we can identify some of the situations in which prehistoric activities took place. In most instances, the tools were lost or abandoned in areas unrelated to their use and were then subject to repositioning by the weathering actions that generally reshape the landscape.

But in rare and fortunate cases, they remain concentrated at the scene of their actual employment with their relationships relatively undisturbed. A number of such examples have been discovered at Olduvai Gorge and east of Lake Turkana; perhaps the most instructive find was at Koobi Fora in the East Turkana area. There, a concentration of flake tools and pebble choppers occurs in the sediments of what had once been a river delta, distributed among the bones of an extinct hippopotamus. The scatter and positioning of the hippo bones strongly suggests that it was butchered and eaten on the site, and furthermore that the stone tools had been used to do the butchering.

Careful geological analysis at that site and in the surrounding area has enabled us to build a picture of what things were like, not only at that time, but back to more than 3 million years as well, and the picture is not much different from what can be seen in much of East Africa today. Large lakes lay along the north-south fault axes of the rift valleys, much as do Lakes Turkana, Manyara, Eyasi, and others today. These were surrounded by grassy savanna lands, which in turn were crossed by meandering watercourses that ended in deltas and swamps at the lake margins. Riverine and gallery forests were distributed along the banks of the lakes and streams. The evidence suggests that these patches of woodland were the preferred habitats of the earliest hominids, and we can guess that the availability of fruiting trees as both sources of nourishment and refuges from the threat of predators was as important to the Australopithecines as it was to baboons, then as well as now.

The hippopotamus butchery site at Koobi Fora was in the delta of one such stream where it joined the Plio-Pleistocene ancestor of Lake Turkana (formerly Lake Rudolf). There, approximately 2 million years ago, a band of Australopithecines encountered a hippopotamus that had died and been washed downstream, to become mired on its side in the shallow waters of the delta. This then became the scene of an Australopithecine banquet. Hippopotamus hide, however, is pretty well impervious to even the most powerful of modern and presumably ancient carnivores. The jaws and teeth of lions and hyenas cannot penetrate it until putrefaction has proceeded to a considerable extent, and we can surmise that the teeth and fingernails of the Australopithecines were not adequate for the task. But modern experiments have shown that Oldowan tools, even in the much weaker hands of twentieth-century *Homo sapiens*, will suffice to gain access to the meat which is beneath the toughest of animal skin. It would seem, then, that the possession of chipped stone tools allowed the Australopithecines to scavenge on material that no other potential user could get at, and it is just possible that the early hominids were able to exploit a niche as pachyderm scavengers for at least a part of their subsistence. At the Hippopotamus/Artifact site at Koobi Fora, they evidently made their tools on the spot and feasted on the exposed upper side of the animal. The other side remained buried in the mud and silt, which preserved it down to the smallest toe bones until both it and the scatter of stone tools were excavated by Richard Leakey's team some 2 million years later.

At Olduvai Gorge, the careful analysis by Mary Leakey, Richard's mother, has shown that Australopithecine tools were associated with faunal remains from mice on up to elephants, although the most common bones are those of antelopes and smaller mammals. And an analysis of the surface of many of the bones with the use of a scanning electron microscope, comparing the scratch marks with experimentally treated samples, has provided conclusive evidence that stone tools were being used to butcher a good number of the animals whose bones have been recovered.

Modern chimpanzees and baboons will kill and eat young gazelles, bush pigs, monkeys, and other animals when the occasion arises, but it is principally an opportunistic and relatively uncommon adjunct to their normal way of life. The Plio-Pleistocene Australopithecines, however, were evidently more systematic in their use of animals as dietary resources. We can guess that they started from a kind of chimpanzee-like opportunism, starting with relatively helpless newborn prey. Gradually they extended their pursuit to just slightly older juveniles, and eventually they developed the skills and capabilities that transformed them into full-scale hunters and proper members of the genus *Homo*.

The archaeological sequence from Bed I to the overlying Bed II at Olduvai provides the best evidence we have for that transformation. The tools at the lower levels of Bed I are crude, and the animal bones with which they are associated are generally from small creatures. By the middle of Bed II, however, tool categories display regular differentiation, and it is clear that the adults of large-sized game animals were being regularly hunted. What we can see there, apparently, is the record of the transformation of a precocious bipedal ape into a full-fleged hunting hominid. The arguments about how much hunting and how much scavenging they did continue unabated, but certainly by Bed II times at Olduvai, the early hominids were pursuing a life way that was radically different from that of any other Primate known. It is tempting to associate this change in life way with the development of those traits in which modern humans are most different from their closest nonhuman relatives.

THE EARLIER ORIGIN OF TOOL USE

We can see clearly, then, that the bipedalism and related dependence upon hand-held tools of the early Australopithecines preceded the focus on hunting behavior by at least a million years. And, obviously, the tools being used at the earlier stages were made of perishable materials. Although we can never "know" this, we can guess that the wielding of a pointed stick was the crucial element which led to the change in selective forces that produced a tool-dependent biped in the first place. As the American anthropologist S. L. Washburn has pointed out, the addition of a simple digging stick to the behavioral repertoire of a baboon could nearly double its food-getting efficiency. This, then, may very well have been the key that allowed the Australopithecines to compete successfully with baboons and wart hogs for one of the basic items of subsistence available on the savannas.

To this we could add the suggestion that the digging stick redirected is a more effective defensive weapon than even the formidable canine teeth of the average male baboon. After all, to bring canine teeth into effective use, the baboon literally has to come to grips with its adversary, and if that happens to be 200 pounds of hungry leopard, the chances are poor that even the most powerful baboon can get away unscathed. If the leopard is fended off by a five-foot length of stout pointed stick, however, the survival chances are slightly better. And if the leopard should throw caution to the winds and charge, the butt of the stick can be planted on the ground with the point directed toward the oncoming predator, which would literally impale itself.

As a dual-purpose defensive weapon and food-getting device, the digging stick would have given the earliest hominids a good competitive edge over the baboons, with whom they shared the Pliocene savannas of East Africa. It is certainly plausible to suggest that these were the circumstances that led to the shaping of the ingrained tool-dependence which is the closest thing to instinctive behavior that we possess. Along

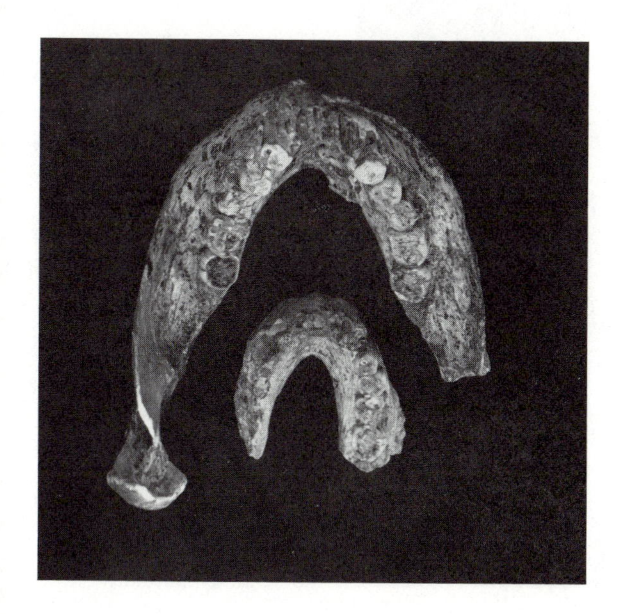

Australopithecine sexual dimorphism, as illustrated by a large (SK12) and a small (SK74a) mandible from Swartkrans in the Transvaal region of South Africa. The difference in robustness greatly exceeds the greatest male-female difference one would normally encounter in a modern human population.

with this, we would expect the development of a mode of locomotion that freed the hands for a tool-wielding role. Indeed, the essentials of hominid bipedalism were already visible in Lucy over 3 million years ago at Hadar.

We could raise the objection that even wielding a stout pointed stick, a three- to three-and-a-half foot tall Australopithecine would be quickly tumbled by a smack or two from a leopard's paw. But Lucy was female, and if the males of her group were double her size and robustness, then the scenario becomes more plausible. Although no skeleton at that date is anywhere near so complete as Lucy's, mandibles both at Hadar (Afar) and at Laetoli show that sexual dimorphism was quite as marked as it is in the other terrestrial primates. Even the canines and lower first premolars show an elongation that suggests a pongid condition in their not-too-distant past.

If the earliest Australopithecines were tool-using bipeds, there is not much reason to expect that their other adaptations were much different from those of nonhuman terrestrial primates. With molar teeth closer to gorilloid than chimpanzee size, they obviously depended for their nourishment on the tough seeds, nuts, roots, fruits, and vegetables that could be gleaned from the savanna and its riverine forests. The yearly round of events and their dietary elements must have been quite similar to those aspects of the life of the baboons with whom they shared the area. For that reason, we can suspect that many aspects of their behavior may have been more like that of baboons than what we think of as characteristically human. We can guess that adult males maintained a dominance hierarchy

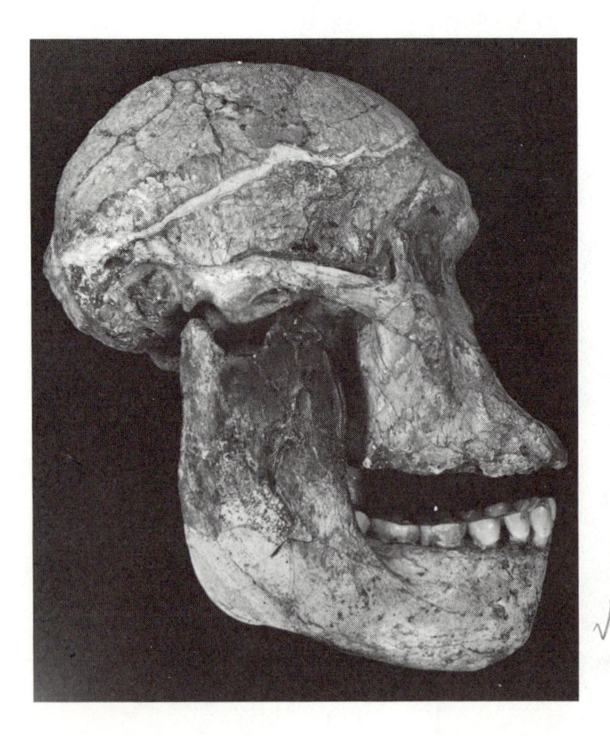

The skull of Broom's "Mrs. Ples," Sts 5, from Sterkfontein, with the jaw SK 23 from Swartkrans. Individuals from these two sites in the Transvaal of South Africa have been considered specifically or even generically distinct, but it is clear that if Mrs. Ples had not lost her teeth, she would have required a mandible of almost exactly the size represented by SK 23. The supposedly robust and gracile South African Australopithecines are very much closer in size than is commonly reported.

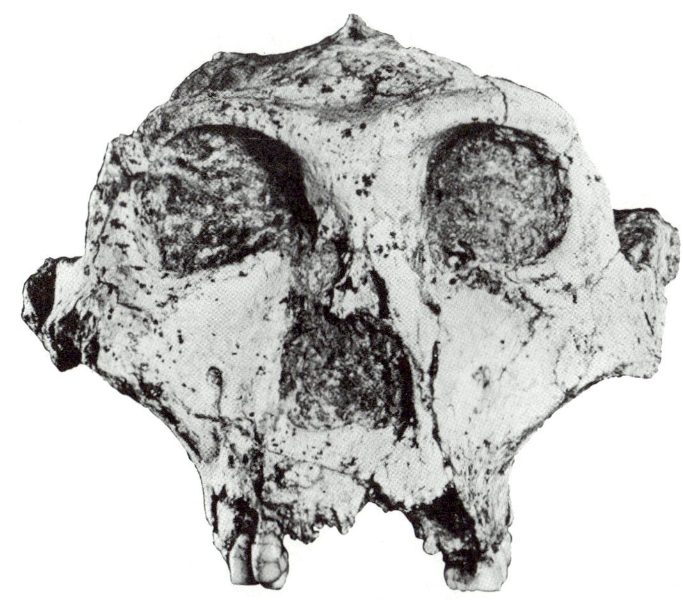

A late Australopithecine of the robust South African type from Swartkrans in the Transvaal. (Courtesy of the Department Library Services, American Museum of Natural History.)

Location of Australopithecine sites: 1. Taung. 2. Sterkfontein. 3. Swartkrans. 4. Kromdraai. 5. Makapansgat. 6. Laetoli. 7. Olduvai Gorge. 8. East Turkana. 9. Omo. 10. Hadar (Afar Depression).

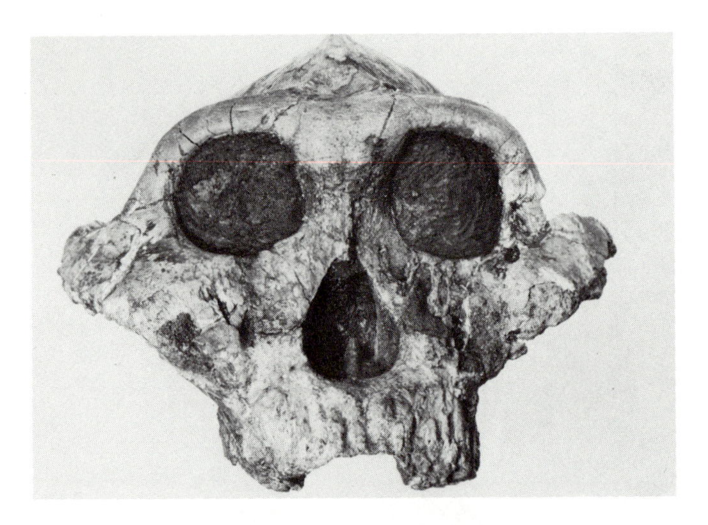

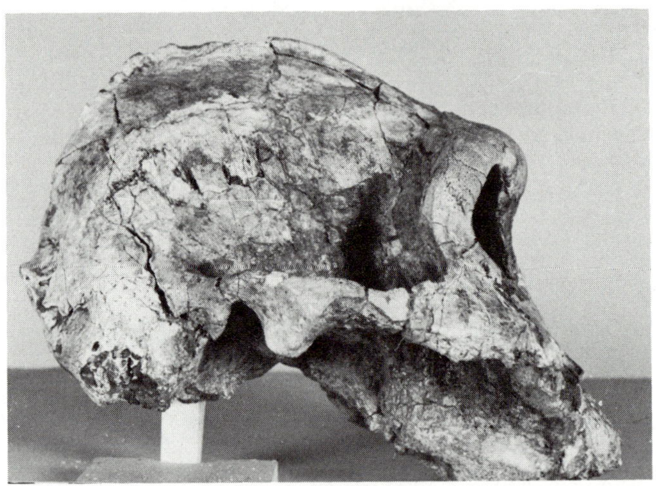

A late robust Australopithecine, ER 406 (above), and a Pithecanthropine, ER 3733 (on facing page), both from the area east of Lake Turkana at about 1.5 million years ago. Note the expanded cheek bones for chewing muscle attachment in the Australopithecine, and the higher and broader braincase in the Pithecanthropine. (Courtesy of the National Museum of Kenya.)

that helped to promote group cohesion and group defense and that included the sexual control of many females by one or a few males. There is no reason to suggest that there was a prolonged male-female bond approaching our concept of monogamy, nor is there any reason to suspect that sexual activity had become a year-round rather than a seasonal enterprise. And there is no reason to believe that they had yet lost the normal primate fur coat.

All told, the earliest hominids might strike us as more ape-like than human, despite their tools and their gait. It is this realization that has led

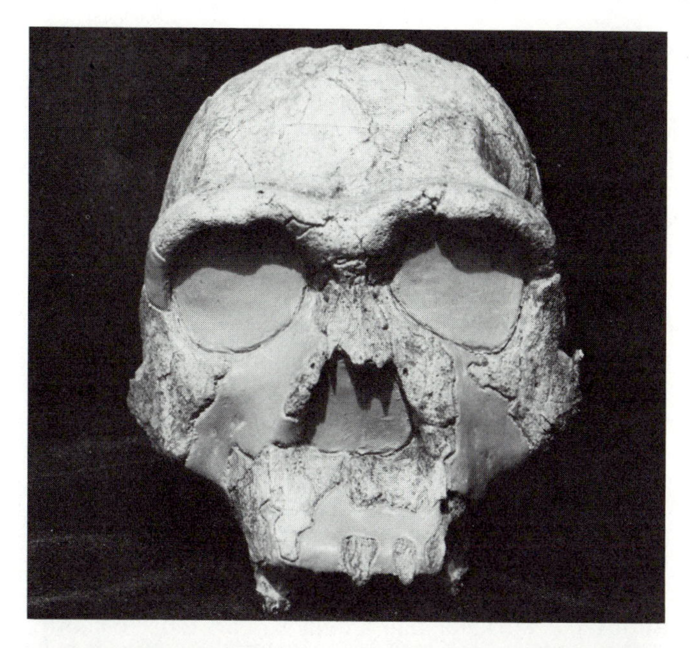

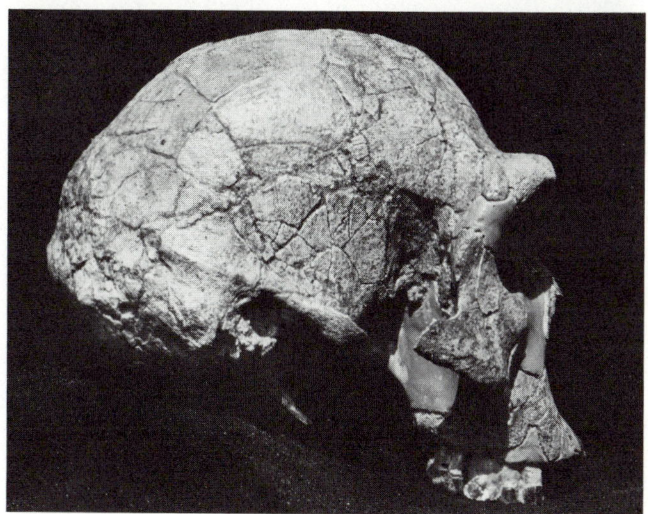

most scholars to grant them generic distinction from true human beings. Backing off from the position I took in the first edition of this book, it seems most in line with the evidence to recognize them as belonging to genus *Australopithecus,* as Dart proposed it in 1925. For the moment, there is no good reason to regard them as specifically distinct from what he described, which would make them, tentatively, *Australopithecus africanus.* Of course, further clarifying discoveries could cast things in quite a different perspective.

HUNTING AND NICHE DIVERGENCE

If systematic hunting was not a significant aspect of early hominid behavior for at least a million years that we know of (and possibly more about which we have no evidence as yet), it eventually became so, as the archaeological record in Olduvai Gorge shows so graphically in the cultural developments between Bed I and the middle of Bed II. But also at Olduvai and even more clearly at East Turkana, it is apparent that not all Australopithecines added a hunting component to their subsistence behavior.

The record shows butchering technology growing in sophistication through time, paralleled by increasing evidence for the successful hunting of large game animals. The record also shows that brain size was increasing and molar tooth size was reducing in one hominid line. But further evidence shows that brain size did not increase in another hominid line while molar tooth size did. The Leakeys' famous "Zinj" find of 1959 has molar teeth of fully gorilloid size. Its brain is also of gorilloid size, which is to say that it is only half Pithecanthropine size and merely one-third that of the modern average.

"Zinj" at 1.75 million years is very nearly duplicated by the ER 406 skull from Ileret, east of Lake Turkana, half a million years more recently, by which time there is also evidence for a fully emerged Pithecanthropine—for example, ER 3733 from Koobi Fora. We are constrained to accept a relationship between the evidence for increasingly successful hunting and cerebral expansion, and if that is the case, then the Australopithecines that continued on through the Lower Pleistocene *without* showing an increase in brain size were probably not engaged in systematic hunting activity. The increase in molar size to a fully gorilloid level suggests a concentration on plant food.

It would appear, then, that starting with the evidence for the first stone tools at the Plio-Pleistocene boundary, a division arose in hominid subsistence strategies between those that concentrated more on hunting and those that focused on the products of the Plant Kingdom. The former became transformed into what we recognize as belonging in genus *Homo*, while the latter remained as members of genus *Australopithecus*. Since the clearest and best-described example of the surviving line of robust Australopithecines was the Leakeys' *"Zinjanthropus" boisei*, and since most feel that the generic designation is unwarranted, there is a tentative feeling that the robust lineage can be identified as *Australopithecus boisei*.

At the moment we have no idea how the separation between the robust line and the lineage that evolved into *Homo* took place. Presumably, some kind of ecological event isolated one group from the other, during which time speciation occurred. This has happened often enough to have produced specific distinctions between many related groups of African monkeys, and there is no reason why it could not have occurred in the hominids at least once. Because the Australopithecines were all essentially savanna dwellers, the crucial separation could have occurred during one of the damper periods when the African rain forest extended from the Congo basin all the way across East Africa to join with the rain forest on the east coast, thus separating the savanna into isolated northern and southern regions.

Did the northern Australopithecines develop the rudiments of hunting during that separation and take the first step that propelled them towards becoming *Homo*? Or was it the southern group that did so? Or was it some other sequence of events? We simply have no evidence as yet. All we know is that the hunter was the only one left by the end of the Lower Pleistocene. Did the hunters do in their vegetarian cousins? Certainly that is a possibility. And if that is a form of catastrophism, we cannot deny that such things have happened frequently enough in the fossil record. Extinctions are more common than survivals, and it is no denial of the principles of evolution to show how the successful development of one group spells doom for another.

chapter eleven

The Pithecanthropine Stage

THE CONSEQUENCES OF HUNTING

The obvious skeletal changes that convert an ape into an Australopithecine are the shortening and spreading of the pelvis, accompanied by the modification of the feet for support rather than grasping, and the reduction of the formerly projecting canine teeth. These are related to the adoption of a bipedal mode of locomotion and hand-held weapons as a mode of defense. They are very ancient and fundamental developments and are united in a biobehavioral complex that provides a necessary base for the later development of those largely behavioral traits that we have come to regard as constituting the essentials of human nature.

The obvious anatomical changes that convert an Australopithecine into a Pithecanthropine are even simpler, but the implications are at least as profound. The key element signalling that conversion is the increase in brain size. Australopithecine brain size was approximately 500 cc., which is about average for the larger of the Anthropoid apes. The Pithecanthropine average is roughly twice that, which puts it well within the lower part of the normal range of variation of modern humanity. Surely this must be related to the development of that constellation of behavioral capacities that is distinctly human. Just as surely, it was the selective forces of the nonpongid subsistence strategy adopted by the emerging Pithecanthropines that led to that cerebral expansion and its implied behavioral correlates.

The crucial factor, as a number of observers have noted, was the adoption of a hunting life way. In spite of the casual examples of baboon and chimpanzee predatory activity, the systematic pursuit of animals for food is a profoundly unprimate kind of activity. We could guess, then, that most of those features that make humans unique among the primates—except bipedalism and its correlates—are the results of the retooling that occurred when the genus *Homo* emerged as a major predator early in the Pleistocene. Because this remodelling involved only very minor visible changes from the neck down, our focus on the importance of the changes in brain size and form would seem to be justified.

In fact, the one postcranial change that is noticeable, if minor, appears to be more a direct correlate with brain size than with any change in bodily usage. Bipedal locomotion had been perfected during the Australopithecine stage, although there were minor ways in which the pelvis and the upper end of the femur (thigh bone) differed from modern form. Biomechanical analysis has shown that the greater iliac flare and longer femoral neck of the Australopithecines were actually more efficient and required less muscular effort than their modern counterparts. The modern configuration appears to have developed during the Pithecanthropine stage. It has been suggested that it was created by the development of the wider birth canal that was necessary to bear the larger-brained infants of the genus *Homo*.

Another correlate of brain size increase is a decrease in the male-female body size difference. Sexual dimorphism remained marked in the Pithecanthropines, but it had reduced from its Australopithecine extreme. The reduction in dimorphism was not caused by a decrease in male size and robustness, but rather by an increase in female size. Again, we can suspect that the selective forces that led to this were those related to the bearing of large-brained infants. Not only that, but carrying an infant for a full nine-month term of pregnancy, rather than the shorter pongid

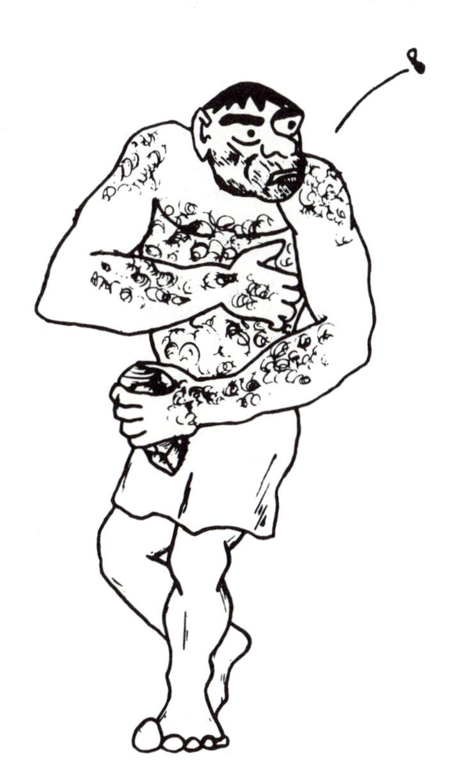

The Pithecanthropine Stage. Note the artist's conception of early man as brutish and flea-bitten, clutching his "hand-axe."

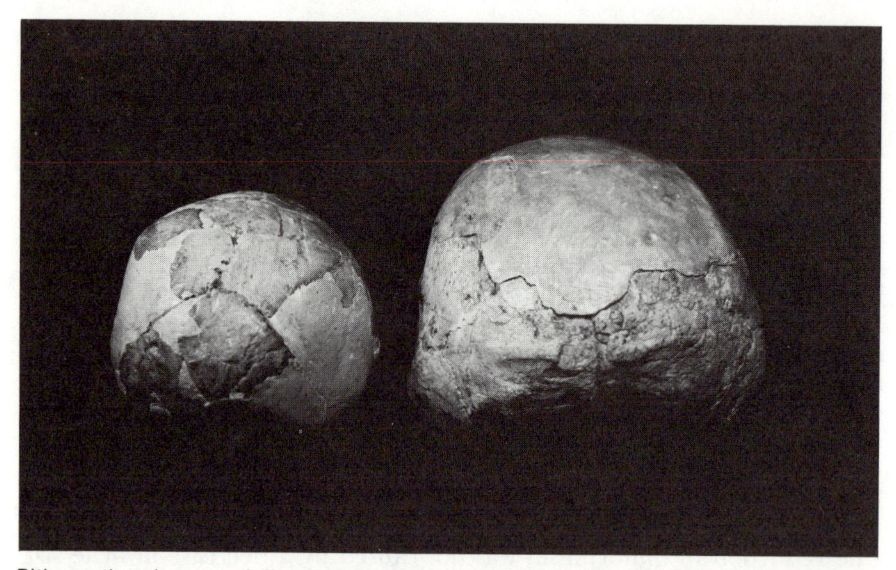

Pithecanthropine sexual dimorphism shown between a female (OH 13) and a male (OH 9) from the middle of Bed II at Olduvai Gorge, Tanzania. The broad base of the male skull provided an attachment area for the muscles of a neck that was very much stouter than that of the female.

term, is more easily accomplished when maternal body size is larger than it had previously been.

Larger brains imply greater intellectual capabilities, and surely that was the key to the successful adoption of hunting as a subsistence strategy by early members of the genus *Homo*. Not only is the brain larger, but an appraisal of the sulcal pattern from the evidence left on the inside of the cranial bones shows that the configuration in the area that controls articulate speech in modern humans is now more modern than pongid. But unlike brain size and conformation, intelligence is not a simple product of heredity. Realized intellectual capacity occurs only after years of trial and experience. Human intellectual achievement is vastly greater than that possible for even the brightest of our pongid relatives, but its mature form is made possible only by a prolonged period of infant and juvenile dependence during which the young are nurtured, protected, and given the benefits of the experience previously acquired by their elders. This period of prolonged dependence is made possible by an equally long period of parental responsibility. And in a situation where hunting activities take males away from the group for days at a time, the role of instructor and protector is better played by females who are considerably larger than the three-and-a-half foot early Australopithecine females.

Although sexual dimorphism was therefore less pronounced than it had been during the Australopithecine Stage, it was still maintained to a greater extent than we currently see in modern human populations. The stress put on the male physique during hunting activities was such that muscularity, joint reinforcements, and general skeletal robustness were all developed to a degree not now encountered. However much stealth and cunning were used in tracking and stalking, the moment inevitably came when the hand-held spear was thrust into the intended victim. The chances that a ton or so of Pleistocene buffalo would quietly and obligingly expire at the first jab of a hunter's spear were small indeed. In the twitching and thrashing of wounded prey, it is certain that during the Early and Middle Pleistocene the hunters regularly got banged around a bit. Torn knee ligaments, broken bones, dislocations, or cracked skulls could easily have had fatal consequences. Recently some doubts have been expressed about whether large game hunting was in fact being pursued by the Middle Pleistocene hominids, but as the sportsmen of the Middle Ages and more recently were well aware, even a wild boar can be a formidable adversary indeed. Even if "large" game were not the focus of the Pleistocene hunters, pork was definitely on the menu, and the bacon was brought home with a far more rudimentary weaponry even than that of the Medieval boar chasers. Not surprisingly, we see the development of bony and muscular reinforcements in the skeletons of the male hunting hominids. Skull walls and long bone shafts are thicker than before or since, joints are expanded and reinforced, and the muscle markings suggest great bodily strength.

Actually, we are more than a little shy of good evidence for the nature of the postcranial skeleton of the Pithecanthropines, but a recent if immature find from West Turkana in 1984 has given us just a bit more

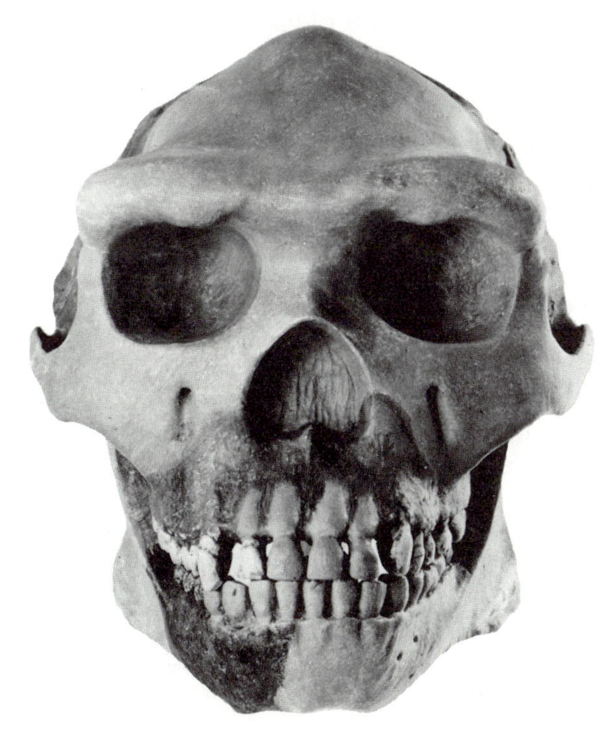

Pithecanthropus IV, as reconstructed by Dr. Franz Weidenreich. The dark parts are original and the light, reconstruction. The whole back part (not visible) was preserved. Recent finds in Java and China have confirmed the accuracy of Weidenreich's reconstruction. The individual was probably a male. (Courtesy of the Department Library Services, American Museum of Natural History.)

to go on. This specimen, WT 15,000, was discovered by Kimoya Kimeu, one of the most successful field workers associated with the Leakey family enterprises, late in the field season of 1984. The discovery was that of a nearly complete skeleton of a 12-year-old boy in strata that can be dated to 1.6 million years ago. Brain size is 900 cc., which is just what we would expect for a Pithecanthropine. Stature was 1.68 meters, which is just over 5 feet 5 inches, and the indications are that, had he grown to adulthood, he would have been a robust 6 feet tall.

The most complete picture of the Pithecanthropine Stage is based upon the fragments excavated from Zhoukoudian, (Choukoutien), just southwest of Beijing (Peking), China, between the late 1920s and the beginning of World War II in the Far East. In terms of the Pleistocene glacial sequence, these date to an interglacial time somewhere around 300,000 years ago. They are somewhat more recent than the remains that Dubois, and later von Koenigswald, discovered in Java, but they are clearly the same sort of thing.

Aside from the expansion of the brain, the other major contrast between the Australopithecines and the Pithecanthropines was in the dentition and its supporting facial skeleton. Pithecanthropine front teeth were slightly larger, but the molars had reduced to the point that they were within the upper limits of the modern range of variation. With a significant quantity of meat as a regular part of the diet, the amount of chewing formerly necessary was considerably diminished. The amount of mastication necessary to reduce animal protein to digestible form is far less than that for vegetables. Ruminants are forced to chew, chew, and rechew their food so that a thorough mixture with salivary enzymes will

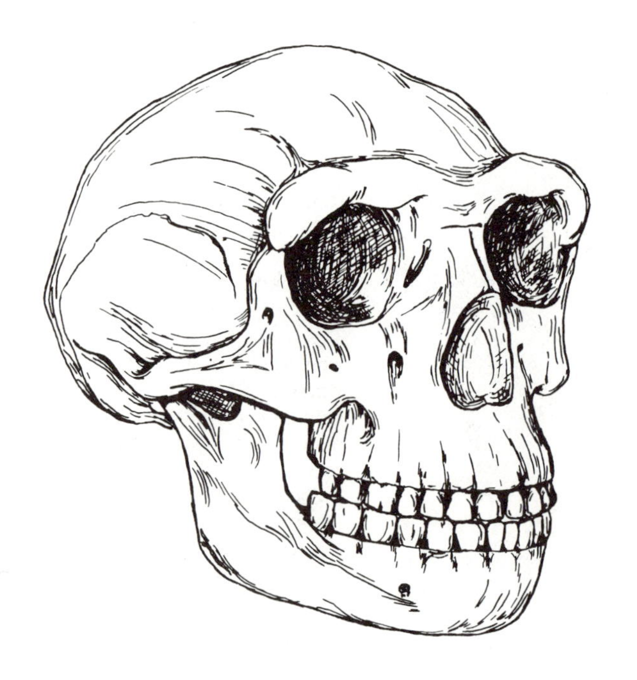

A composite reconstruction made under the direction of Franz Weidenreich and based upon the Pithecanthropine fragments found at Zhoukoudian, near Peking. The specimen was originally called "*Sinanthropus pekinensis.*" The individual on which this was based was almost certainly female.

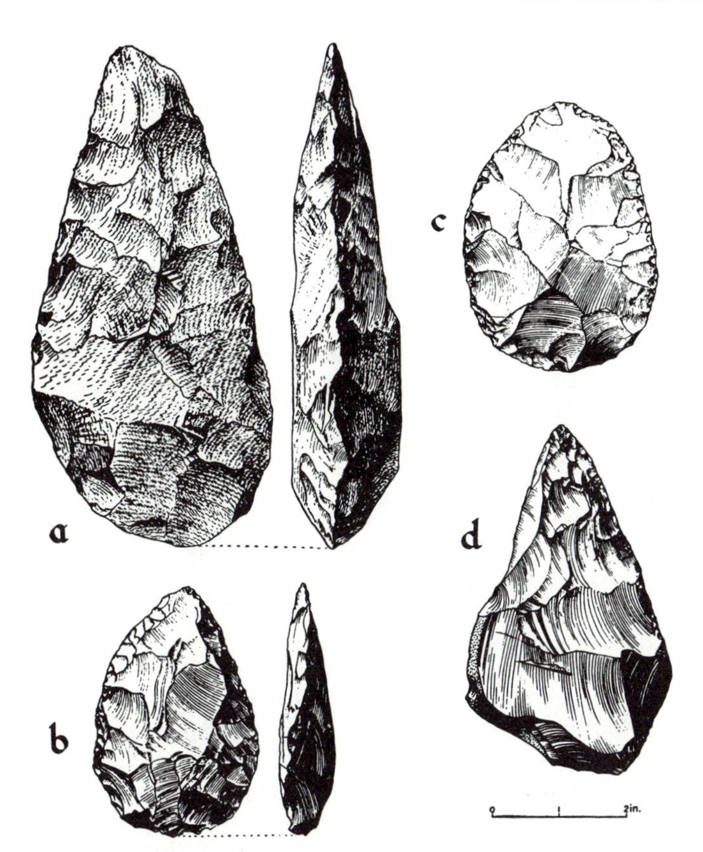

Lower Paleolithic tools. While most of the tools of the Lower Paleolithic are less regularly shaped flakes, picks, cores, and fragments, most assemblages include an occasional representative of the "hand axe" type first recognized by Boucher de Perthes at St. Acheul. Variants on that theme are depicted here: a) Lava biface (hand axe) from Ol Orgesailie, Kenya; b) Biface from St. Acheul, northwestern France; c) Ovate biface from Israel; d) Late Lower Paleolithic biface from Suffolk, England. (Reprinted from Oakley 1950, by permission of the Trustees of the British Museum of Natural History.)

assure its digestibility. A quick look at the molars of a cat, however, will demonstrate that shearing rather than crushing is their main function. For a carnivore, the main purpose of the molars is to reduce food to swallowable size, since animal protein does not require extensive salivary enzyme action before it can start to be digested in the stomach. If Pithecanthropines were eating significantly greater quantities of meat than their Australopithecine ancestors, then they should have had less need for the greater crushing molars of the hominids at the Plio-Pleistocene boundary. With molar size free to vary, then the probable mutation effect could do its work. The result was that reduction took place.

So far we have mentioned the use of cunning and stealth in hunting, but it seems likely that another basic approach was also used. The human adaptation to long-distance locomotion is actually fairly remarkable in its own right, although, considering the flabby physique of the average reader in his armchair, it is perhaps less easy to appreciate this than it should be. As an example of the potential capability of a well-conditioned human being, we can cite the mode of hunting practiced even today by certain peoples. This involves literally walking one's quarry into the ground. South African Bushmen, American Indians, and Australian Aborigines are noted for this simple, if rather exhausting, technique. The hunter takes up the trail of a large herbivore, sets out after it at a persistent trot, and keeps it moving until, out of sheer fatigue, it can go no further, at which point the hunter moves in and dispatches it. Because the process may actually drag out over a number of days, it involves a skill in tracking and a degree of patience and endurance difficult for the beneficiaries of the technology of this mechanized age to conceive.

Two physiological facts aid the hunter in this form of activity. First, a large herbivore depends upon the ingestion of great quantities of food of relatively low nutritive value, which means that it has to spend a considerable portion of its lifetime eating. The hunter, on the other hand, fortified by occasional nibbles of concentrated nourishment in the form of dried meat products, nuts, and other high-quality edibles, can keep pushing on without stopping, prodding his quarry along just fast enough so that it does not have an adequate chance to stop and replenish itself. In a couple of days, the hunter's patience is rewarded.

An interesting comment can also be made at this point. Human digestive physiology is quite different from that of the average nonhuman Primate. The normal primate eating pattern is that of the nearly nonstop snack, which is paralleled by nearly nonstop elimination. Admittedly, the fast-food industry in the Western world has shown how easy it is to reintroduce the old primate eating habits, but fortunately this has not been accompanied by a resumption of nonstop feces production. It would seem that the retooling job that made a Primate into a hunter by the Middle Pleistocene successfully accomplished a permanent change in its digestive physiology.

The second physiological fact involves the human ability to dissipate metabolically generated heat. A human being, whose hairless skin is richly endowed with sweat glands, can continue to function effectively throughout the heat of the day. The South African Bushmen capitalize on this fact by running down large quadrupeds in the middle of the day, when the animals are prone to develop heat exhaustion if they attempt any continued rapid locomotion. In the tropics, mammalian life usually reposes in the shade during the hot part of the day. It is not without significance that all of the predatory carnivores that survive as the result of active pursuit of prey engage in their maximum expenditure of energy in the relative coolness of the early morning or late afternoon and evening.

The relative night blindness that humans have inherited from their diurnal primate ancestors means that they could not even begin to mount an effective challenge to the established predators at the time when they are most active. But there was no competition at all for the role of a diurnal predator, and this was the niche that was adopted by the early members of the genus *Homo*. Except for mad dogs, man alone goes out in the noonday sun—nor is this solely a peculiarity of the English, either. It would appear that the development of human predation long ago capitalized on the limitations a coat of fur placed upon mammalian activity during the heat of the tropical day, and we can suspect that the perfection of the hominid pelvis for long-distance walking was accompanied by the effective loss of body hair.

At the same time, the intensity of ultraviolet radiation poses something of a problem to the hairless tropic-dweller, since it greatly increases the chances of developing skin cancer. The solution is the development of a concentration of the protective pigment melanin. To follow up the train of these observations with a further speculation, it is possible to postulate that with the development of effective hunting techniques, somewhere in between the Australopithecine and the Pithecanthropine Stages, people became hairless and black. Later, we account for the depigmentation that occurred in the background of some of the world's peoples, but at present it is sufficient to suggest that all of humankind passed through a heavily pigmented stage.

PITHECANTHROPINE CLASSIFICATION

Yet to be considered is the formal taxonomic designation of the Pithecanthropines and their geographic distribution. The controversy that raged over the status of the original Pithecanthropus around the turn of the century now belongs to history, although one or two professional scholars would still like to see it recognized as representative of a sideline that became extinct without issue. Although a great many anthropologists balk at accepting the Australopithecines for one reason or another, many are willing to regard the Pithecanthropines as genuine human beings who were ancestral to all later forms of humanity. Some would still prefer to posit an ancient form of "true man," as yet unfound, but there is little doubt that the Pithecanthropines can be considered genuine, if primitive, human beings. As a result, the validity of the term *"Pithecanthropus"* as a formal generic designation has been questioned, and most authorities now regard them as belonging within genus *Homo*. The original specific designation is still considered tentatively valid, and many workers are quite happy to refer to the Pithecanthropines as *Homo erectus*.

THE PITHECANTHROPINE SPREAD

A generation ago, when Pithecanthropines were unchallenged as the oldest hominids known, and when they were recognized only in Java and China, there was a general feeling that Asia had been the cradle of mankind. Now, however, with Africa possessing the strongest claims to be the initial human homeland, there is much more willingness to view the Pithecanthropines as having spread throughout the area that the archaeological remains indicate was inhabited, rather than to regard them as having been restricted to one small province. New finds and the reappraisal of some old ones confirm this suspicion. For example, Robinson reappraised the Swartkrans specimen which he and Broom originally called "*Telanthropus*" and demonstrated that it could be considered a proper Pithecanthropine.

This was the first evidence discovered that showed the contemporaneity of an Australopithecine with an early Pithecanthropine. Further such evidence, although tantalizingly skimpy, has been found at Olduvai Gorge, and it has finally been clearly and dramatically confirmed by the spectacular East Turkana finds made during the last decade.

One problem with trying to identify possible Pithecanthropines on the basis of jaws and teeth and not much more is that they are indistinguishable from Neanderthal jaws and teeth. With no more skeletal evidence available, which stage gets assigned is determined by dating. If the specimen is Middle Pleistocene in age, jaws and teeth of Pithecanthropine size can be comfortably referred to the Pithecanthropine Stage. An example is the famous Heidelberg mandible of 1907. The teeth correspond quite nicely to those discovered at Zhoukoudian, and because the dating can be shown to be approximately the same, it is reasonable to infer that Heidelberg represents the northwesternmost extreme of the Pithecanthropine range, just as *"Telanthropus"* may represent the southwesternmost extreme. Joining Heidelberg as European representatives of the Pithecanthropine Stage are specimens from Germany (Bilzingsleben), France (Arago), and Greece (Petralona). In Africa, the Kabwe ("Rhodesian") skull from Zambia, the Saldanha skull from near the Cape of Good Hope, and Bodo from Ethiopia, as well as Ternifine from Algeria, show the form as well as the distribution of the African Pithecanthropines.

The Pithecanthropines at Olduvai and in the East Turkana area can be associated with a tool-making tradition that had developed straight out of the earlier Oldowan; they also occur in deposits that can be dated by

Sites of major Pithecanthropine discoveries. 1. Java. 2. Zhoukoudian. 3. Lantian. 4. Vertesszöllös (Hungary). 5. Heidelberg. 6. Arago (France). 7. Rabat (Morocco). 8. Ternifine (Algiers). 9. Bodo (Ethiopia). 10. East Turkana. 11. Olduvai Gorge. 12. Swartkrans.

Pithecanthropine skullcap from site LLK in Bed II of Olduvai Gorge, Tanzania. (Photo by Dr. L. S. B. Leakey; copyright National Geographic Society.)

radiometric means. At about 1.5 million years old, they are older than almost all other Pithecanthropine remains. Occasionally there have been claims for a comparable degree of antiquity in Java, but the evidence is shaky at best, and most observers now feel that the Pithecanthropines in Southeast Asia do not go back as far as the Lower Pleistocene. All of the more northerly representatives, including the Chinese and the increasing number of European discoveries, are no more than half that age, belonging in the Middle Pleistocene, which began only 700,000 years ago.

According to reliable evidence, the Australopithecines were confined to the continent of Africa. Presumably, the Pithecanthropines evolved out of them there and only later spread out of their continent of origin. If the Javanese dates can be trusted, the spread to the east occurred more rapidly at an equatorial latitude, but the spread to the north, whether northwest (Europe) or northeast (China) took much longer to accomplish. Lower Paleolithic tools occur in a continuous distribution from Africa through India to Southeast Asia during the Middle Pleistocene, and wherever human skeletal remains are found with them, these are of Pithecanthropine form. Just recently, the find of hominid cranial fragments associated with the extensive Lower Paleolithic archaeological remains in the Narmada Valley in India has reinforced this point.

Physiologically, people are still tropical mammals, and that must have been no less true for the early Pithecanthropines. Presumably after their subsistence pattern was developed in Africa, they were easily able to fill the diurnal hunting niche all the way across the tropics of the Old World. The spread north into what we curiously refer to as the "temperate" zone must have been possible only with the development of cultural adaptations that could compensate for the physiological limitations imposed by the tropical primate heritage. The development of clothing, of course, is one such compensation. So is fire. There has been some tentative suggestion that Pithecanthropines in both Europe and China were found with traces of charcoal, but there is serious doubt about whether that was the product

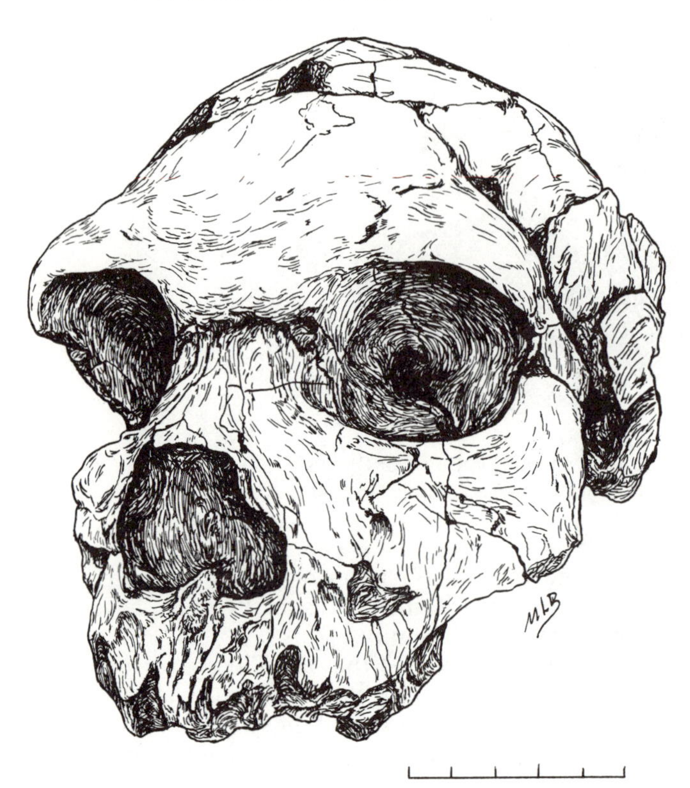

Bodo, an *erectus* skull from Ethiopia. (Drawn from a cast by M. L. Brace; courtesy of Professor T. D. White, University of California, Berkeley.)

of human activity or simply the result of naturally occurring brush fires. Human beings eventually did control the use of fire, and it is tempting to suggest that this is what finally allowed the permanent settlement of the northern areas of habitation in the Old World at the end of the Middle Pleistocene. We shall return to this matter later on.

The initial occupation of previously uninhabited areas, whether the tropics of the Old World, or, later, the temperate zone, was obviously in the nature of an actual movement of people—even if this only involved the excess population of locally established groups budding off and inhabiting the next territory just a few miles away. However, once the habitable world was occupied, development from the Pithecanthropine Stage to the succeeding ones probably occurred gradually and simultaneously throughout the entire occupied world. The traditional interpretation has tried to view them as arising at one point, after which they would presumably have spread by extinguishing the conservative local inhabitants, wherever they might be. But since the circumstances that led to their development were widely distributed, it seems more in line with evolutionary mechanics to expect them to have arisen throughout the broad zone of hominid habitation rather than at any one restricted spot.

Subsequent migration to previously uninhabited areas would be the next expectable step. After the initial Pithecanthropine spread, invasions of any note probably did not occur until the time of the great population imbalances and technological disparities that grew out of the food-producing revolution within the last 10,000 years.

The reason for this view becomes clear if we appraise the nature of the cultural adaptive mechanism. Cultural adaptations can and do diffuse with ease across the boundaries of specific cultures. The bow and arrow, for instance, diffused to most of the corners of the globe in a relatively short period of time, and the documented spread of the use of tobacco indicates a rate of diffusion and a disregard for cultural boundaries that is nothing less than phenomenal. With the high degree of mobility and relative cultural uniformity functionally characteristic of the Lower Paleolithic hunters, any significant advance in hunting technique, food preservation process, or the like must have diffused quite rapidly throughout the inhabited world—accompanied by the inevitable, if not large, leakage of genes across population boundaries as well. With the major forces shaping human evolution heavily influenced by major cultural adaptations,

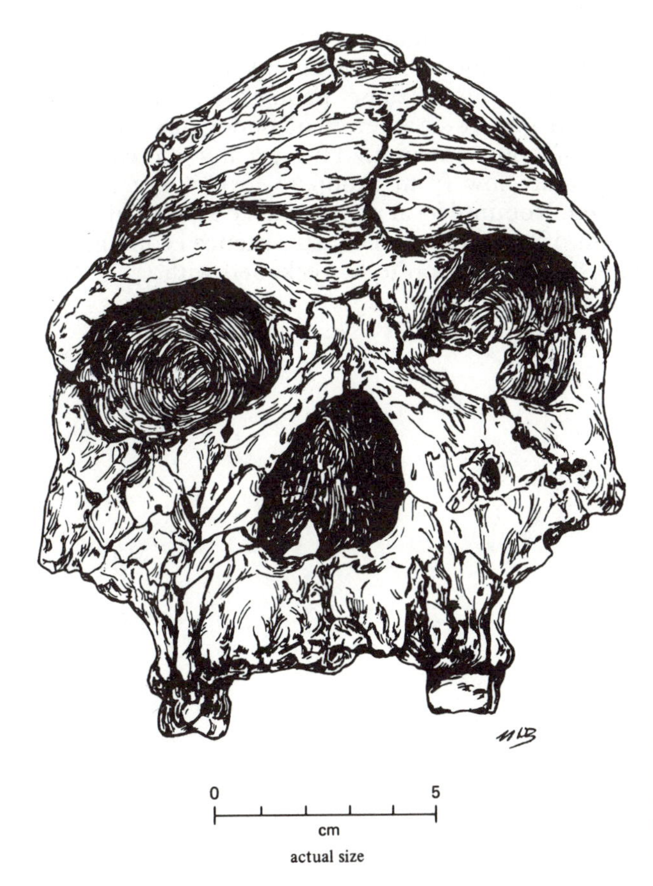

Arago XXI. (Drawn by M. L. Brace from a photograph taken with the permission of Drs. Henry and Marie-Antoinette de Lumley.)

and with the latter effectively diffused, whatever their local origins, we can postulate relative similarity in the selective forces operating on human form throughout the Lower Paleolithic. Similar forces would have produced similar evolutionary consequences in widely separate areas, even without an accompanying slow genetic interchange, although this must have occurred as well.

Starting with the Pithecanthropines, it is just possible that the diffusion of the cultural reasons for the specific physical changes which characterize the succeeding stages of human evolution were rapid enough so that the unity of the human species was maintained at any one time. Development from one stage to the next, then, would have proceeded at approximately the same time and at roughly the same rate throughout the inhabited world. The probability that a given population will be genetically more like its precursors in the same locality is, of course, greater than the probability that it will be genetically closer to groups in adjacent areas, and this allows for the development of regional peculiarities; however, at the same time, genetic material is continually being exchanged with adjacent areas. The result is that no human population has ever become different enough from its contemporaries to warrant formal taxonomic recognition.

Once again, observations on contemporary representatives of a hunting and gathering form of subsistence economy provide evidence that reinforces this suspicion. Among these people, *exogamy*—seeking mates from other unrelated groups—rather than endogamy—inbreeding—is a virtually universal phenomenon. This greatly increases the possibilities not only for gene flow from group to group, but also for information transfer as well. Not until local populations became sedentary, following the development of a food-producing subsistence economy, did group endogamy become a phenomenon to be reckoned with.

chapter twelve

The Pithecanthropine to Neanderthal Transition

THE "RHODESIAN" FIND

If the Neanderthals were characterized by the development of a Modern-sized brain in a creature that still had a Middle Pleistocene body and face, then their predecessors late in the middle part of the Pleistocene ought to have shown signs that final expansion of the brain was taking place. This indeed is exactly what we find. The first of these late Middle Pleistocene finds to be recognized was the famous "Rhodesian" skull, discovered in 1921 deep in a mine shaft in Kabwe (then Broken Hill) in Zambia (then Northern Rhodesia). Brain size, at just over 1,250 cc., is well above the Pithecanthropine average of 1,000 cc., but below the modern

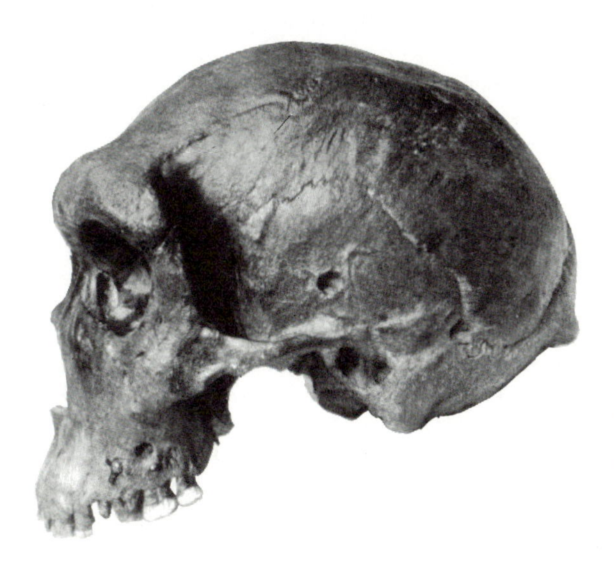

The Rhodesian skull, an African Neanderthal. (Courtesy of the Department Library Services, American Museum of Natural History.)

male mean of 1,450 cc., although well within the normal modern range of variation. The face is a good, unreduced Middle Pleistocene representative, capped by a stupendous brow ridge. The neck muscle attachments likewise indicate that the head was set upon the body of a powerfully constructed Middle Pleistocene male.

The date of the Rhodesian skull has been a matter for debate ever since it was found. Recently, a new technique based upon the racemization of aspartic acid has been applied, and this yields a date of 120,000 years, late in the Middle Pleistocene. But there are many doubts and questions about the use of such techniques to assess the antiquity of terrestrial as opposed to oceanic material, so there is still no final word on the date of the Rhodesian specimen. All we can say about it is that it appears to be on the boundary between Pithecanthropine and Neanderthal form.

SALDANHA

In 1953, a similar skull vault, lacking the face, was found at Saldanha Bay near Hopefield, a scant hundred miles north of the Cape of Good Hope in South Africa. In the details of its form, it is strikingly similar to the Rhodesian skull. Furthermore, it was found associated with faunal remains of characteristic Middle Pleistocene form, and also with artifacts recognized as final Acheulean, which suggest a late Middle Pleistocene date. All of this would be consistent with an interpretation that would regard this specimen as evidence for the transition from *erectus* to *sapiens.* There has been continuing disagreement about which taxon the specimen belongs to and this could very well be the reason why it cannot be precisely placed.

PETRALONA

A near duplicate of the Rhodesian skull was found by a group of cave explorers near the village of Petralona in northern Greece in 1959. The face and brow ridges suggest Pithecanthropine affinities. The cranial capacity, at 1,220 cc., is halfway between the Pithecanthropine and the Neanderthal levels, but, like the Rhodesian specimen itself, this can only be a tentative suggestion at best because there is no means of assigning a date.

SOLO

The Far East has also obligingly yielded some crucial fossils. As with the initial discovery and the confirmation of the Pithecanthropine Stage, Java has played the central role. Starting in 1931, eleven broken and faceless skulls were unearthed on the banks of the Solo River. These recall the Pithecanthropines, on the one hand, and the Neanderthals, on the other. The bones of the cranial vault are thick, the brow ridges and muscle markings are heavy, and the keeling along the midline, together with other details, looks more than faintly Pithecanthropine. However, the cranial capacity is halfway between Pithecanthropine and Neanderthal/Modern levels, and the date is considered to be from just before the

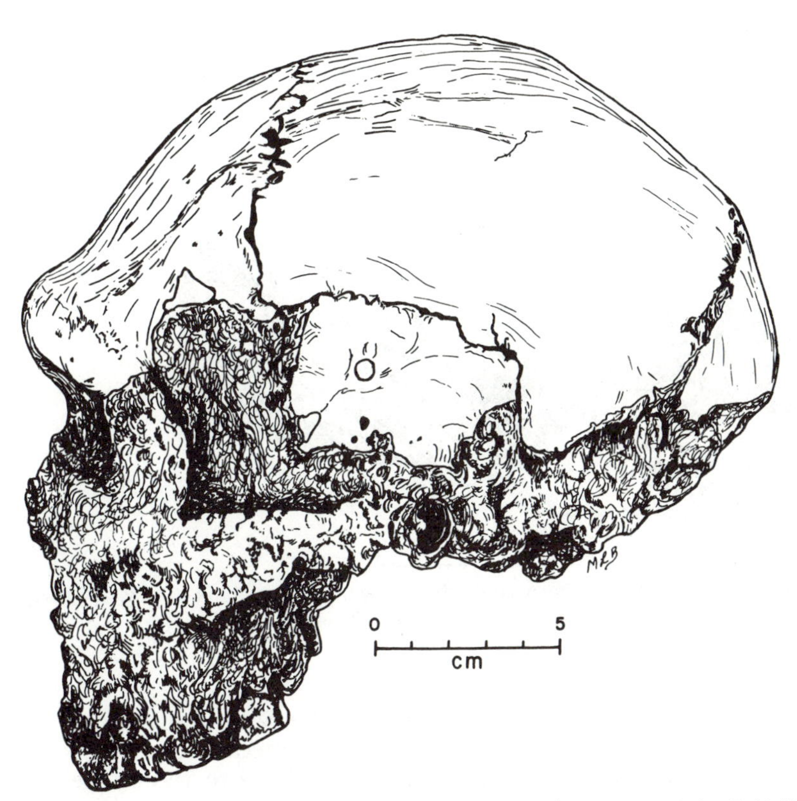

Petralona, an *erectus* skull from Greece. (Drawn from a photograph by P. Kokkoros and A. Kannelis published in *L'Anthropologie,* 1960, and reproduced courtesy of Masson & Cie., Paris.)

One of the Solo skulls, a Javanese representative of the transition between the Pithecanthropine and the Neanderthal Stages. (Courtesy of the Department Library Services, American Museum of Natural History.)

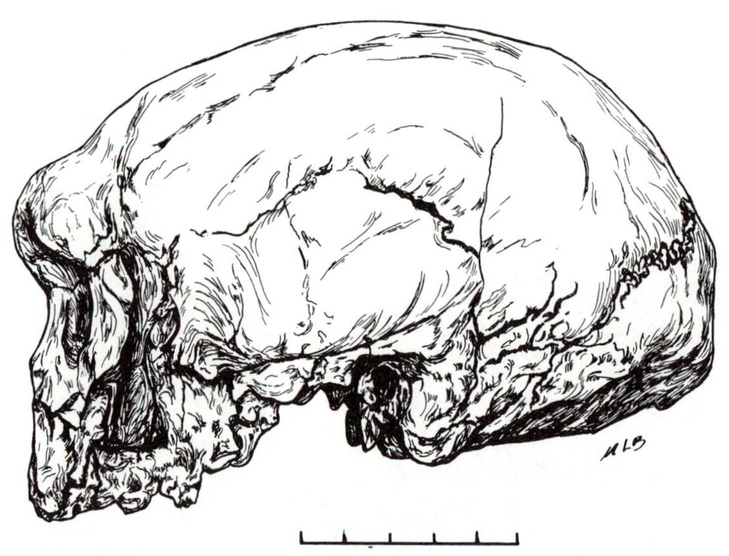

The Dali skull, a Pithecanthropine/Neanderthal transition from China. (Drawn by M. L. Brace from a photograph; courtesy of Professor Wu Rukang.)

beginning of the Upper Pleistocene. All told, the Solo skulls appear to represent an evolutionary transition from the Pithecanthropine to the Neanderthal stage in the Far East. Together with more recent skeletal remains, the evidence is most suggestive that stage-to-stage evolution occurred simultaneously in all parts of the inhabited world: Europe, Africa, Asia, and, by inference, the areas in between.

DALI

The one specimen from the appropriate time level in China also shows all the characteristics of being intermediate between the Pithecanthropine and the Neanderthal Stages. A skull was found in 1978 in river gravels in Dali County, just east of the ancient Chinese capitol of Xian, in Shaanxi. The date is late Middle Pleistocene, and, with a capacity of 1,120 cc., it is appropriately intermediate. Dali is also interesting in that it shows, admittedly in very robust form, some of the same aspects of facial proportion that serve to distinguish modern Asian faces from those in other parts of the world. This adds to our suspicion that modern human form emerged simultaneously in the various geographic regions of the Old World.

SWANSCOMBE

Europe has also yielded specimens from the same time period, and it would seem that there is plenty of evidence to support the uncomplicated view that Neanderthal form emerged gradually from that of the preceding Pithecanthropines. But this is not the way the majority of analysts prefer to look at things, so we should consider the evidence and what is said about it to produce the confusions that pass for current orthodoxy. Three European finds are frequently cited as contradicting this apparently straightforward scheme. These are, in order of their discovery, the specimens from Steinheim in western Germany, Swanscombe in England, and Fontéchevade in France.

The oldest of these "skulls" is the Swanscombe skull. Although there are continuing arguments about just how old it is, it is certainly Middle Pleistocene and possibly in the latter part of the Middle Pleistocene. The pieces of this skull were discovered in a gravel pit of the lower Thames River in southeastern England; the three major fragments constituting the rear of the cranial vault were unearthed in 1935, 1936, and, by an almost impossibly rare piece of good fortune, in 1955. At the time when the initial pieces were found, British anthropologists, because of their long-standing lack of enthusiasm for facing the possibility that modern humanity may have had a Neanderthal ancestor, were desperately eager to find evidence for the existence of "men" of modern form at an earlier time level than that attributable to the Neanderthals. As a result, modern features were stressed whenever possible. Because the all-important facial parts of Swanscombe were missing, opinions concerning its status could be promoted without much risk of encountering solid objections from any quarter whatsoever. By default, then, Swanscombe has been regarded as "Modern" ever since.

Even though the back end of the skull is relatively nondiagnostic in the assessment of the major distinguishing characteristics of evolutionary stage, some features generally accompany those of diagnostic significance, and it is not without interest to discover some of these on the Swanscombe skull. For instance, the greatest width is far back and low down on the skull, and the skull height is remarkably low in proportion to that width. The width across the occipital bone alone is greater than 99 percent of comparable modern skulls, and the bones are remarkably thick. Other indications as well locate the Swanscombe skull right in the middle of the characteristic Middle Pleistocene range of variation, but, lacking the crucial

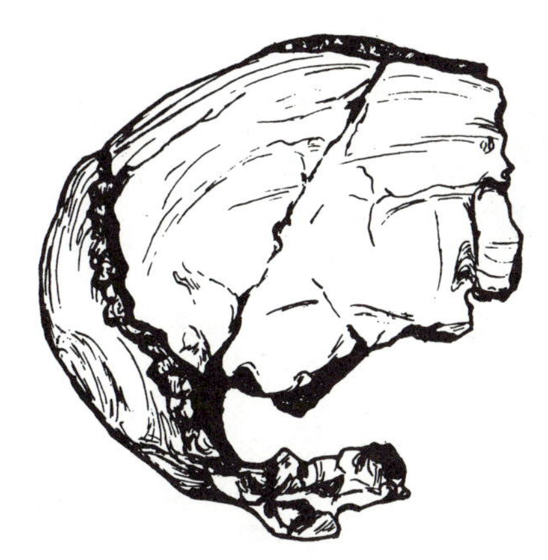

The Swanscombe skull. (Drawn from a cast by M. L. Brace.)

frontal bone and attached facial parts, we are not at liberty to do more than suspect that those parts may well have agreed with the indications of the vault. Unequivocal interpretation of Swanscombe is not possible, but, in marked contrast to the majority of the claims put forward on its behalf, it most certainly provides no evidence whatsoever for the existence of modern human form in the Middle Pleistocene.

STEINHEIM

The Steinheim skull, found two years earlier in a gravel pit near Schiller's birthplace, not far from Stuttgart in west Germany, would at first consideration seem to be a more promising subject for interpretation than the Swanscombe skull. The skull is of about the same antiquity, and it is relatively complete. Much of the face is preserved. Yet the arguments surrounding the attempt to establish its significance show no signs of diminishing. The skull is small and low, and it has a cranial capacity of approximately the Pithecanthropine average. The brow ridge is a formidable bony bar, but the back is rounded and smooth, suggesting modern form. The third molar is markedly reduced. Although the modern form of some of the other parts of the face has also been stressed, there are two principal difficulties in the way of definitive interpretations. First, the whole lower front part of the face is missing, leaving only the molars and one premolar at the rear of the dental arch. This is particularly regrettable because the most crucial features distinguishing Modern from Middle Pleistocene morphology are those centered upon the forward end of the dental arch. The second difficulty lies in the distortion that the skull has undergone. The whole left side of the skull is crushed toward the midline, reducing the width of the base to less than that ever recorded for a normal modern individual (in whom the width of the base tends to be less than for Middle Pleistocene individuals in the first place). The palate has been

The Steinheim skull. (Drawn from a cast by M. L. Brace.)

reduced in width, and the whole of the facial skeleton has been pushed slightly back underneath the skull. As a result of the missing and distorted aspects, it is evident that no unequivocal judgment can be made. Yet, with its small cranial capacity and heavy brow ridge, it can tentatively be regarded as belonging somewhere between the Pithecanthropines and the Neanderthals.

In the case of both Swanscombe and Steinheim, another matter has rarely been given due consideration. Both individuals were female and therefore lacked the bony reinforcements and evidences for heavy muscle attachments that are so prominent on Middle Pleistocene male skulls. In a population in which sexual dimorphism was pronounced, as it was during the middle of the Pleistocene, the choice of any single skull to characterize the appearance of the group as a whole is bound to be misleading. Such would seem to be the case for those who have taken the female form of Steinheim and Swanscombe to show that Modern "man" existed before the Neanderthals.

FONTÉCHEVADE

The other principal fragments for which the cry of "ancient Moderns" has been raised are those found in the cave at Fontéchevade in the Departement of Charente in southwestern France. The discovery of pieces of two human skulls in 1947 was widely hailed by the anthropological world as proof at long last of the existence of Modern form back in the Middle Pleistocene. Closer examination of the circumstances of the discovery and their nature, however, reveals a monumental amount of confusion. As it turns out, one of the skull fragments was not found *in situ* but in the laboratory where the block of material containing it had been brought for dissection at greater leisure. This fragment includes the section of a skull toward the medial part of the left eye socket, between the eyes, and rising a short distance up the forehead. From what we can see, it is apparent that no heavy brow ridge was present, but, also from what we can see, there is no assurance that the fragment came from an adult. It is just as consistent to regard it as a juvenile from a population in which heavy brow ridges developed during adolescence. In any case, it is risky to hang the whole argument for the existence of ancient modern form on one such small and equivocal piece of evidence.

The second Fontéchevade fragment includes the better part of the top of the cranial vault of what was apparently an adult. However, the diagnostic frontal and basal parts are missing, the piece is crumbly, and so much doubt clings to the attempts to project a reconstruction from the available parts that it would be far better to put the finds aside until more complete evidence is discovered. Certainly, at the moment, the only reason for stressing the dubiously "Modern" features of Fontéchevade (or Swanscombe, or Steinheim, for that matter) is the desire that so many authors apparently have to find something less "primitive" that came before the Neanderthals of the last (Würm) glaciation, thereby proving that the latter could not possibly be the ancestors of recent humankind. This desire seems to have its roots in a trend of thinking that becomes alarmed whenever the suggestion is raised that Modern "man" evolved from something less "man-like" than himself.

But if we accept the possibility of evolution in general, and the likelihood of Darwinian mechanics in particular, the hominid fossil evidence from the latter part of the Middle Pleistocene shows just the kind of gradual change from Pithecanthropine status to Neanderthal form that we would expect to find. And it shows the effects of the mediating influence that the increasing roster of cultural phenomena have on the forces of selection that had previously exerted their full effects on human chances for survival.

THE CONTROL OF FIRE

As the Middle Pleistocene came to a close, the control of fire was added to the roster of cultural items, and the northern extent of human habitation was considerably expanded. Because the subject of the human use of fire in prehistoric times is of some importance, it is worth exploring its implications. For one thing, such evidence is a boon to the prehistorian, since it means that the difficulties involved in discovering the remains of ancient habitations are greatly reduced. With the advent of fire, caves were inhabited by humans for the first time—a fact that greatly reduces the number of places the archaeologist has to investigate before getting results. Prior to the advent of fire, caves were studiously avoided at night by prehistoric hominids since they were more in the nature of traps than shelters. The keen visual sense that people inherited from their arboreal precursors, although remarkable among mammals for its acuity of color discernment and depth perception when light is provided, left (and still leaves) them relatively helpless in dim light, and practically disoriented in deep darkness.

Fire is useful in three ways, and is symbolic of a fourth phenomenon of considerable importance. It provides light, which extends the length of time during which a hominid can effectively operate. It provides nocturnal protection, which can convert the limiting confines of a cave into a safe area of refuge. And it provides warmth, which can enable a fundamentally tropical mammal to extend its range into colder climates normally closed to it. The final point—that which fire symbolizes—is related to the reuse of an agreed-upon campsite. Where the depth of deposition in cave sites indicates that they were intermittently used again and again, or even for a succession of days, then it is more than likely that the users were capable of communicating time and place between each other. With this as a possibility, the ability of the group to divide up and agree to meet later at the camp is also a possibility.

We must recognize in this a demonstration of the effective application of the division of labor. Because of the physiological differences between males and females—the latter being charged with the care of infants and young—the most basic form of the division of labor is inevitably by sex. The men are concerned more specifically with the chase, and the women concentrate on vegetable products and slow game. Even such a rudimentary division of labor as this can greatly increase the subsistence base of a foraging group, and its effectiveness is evident in the fact that it is still characteristic of the remaining hunting and gathering peoples. Granting that this is rather a jump from the simple recognition of campsites via reused hearths, it nevertheless seems a legitimate interpretation to offer for a creature for whom specialized hunting activities had begun to play an increasingly important role in group subsistence. Certainly at the end of the Middle and the beginning of the Upper Pleistocene, the evidence shows that there was a considerable increase in the numbers and effectiveness of the prehistoric human hunting populations.

chapter thirteen

The Neanderthal Stage

FINDING AND DATING THE NEANDERTHALS

Of all the human fossils known, Neanderthals have generated the most public interest, serving as the prototype of the cartoon "cave man." Somewhat ironically, now that the reading public has finally gotten to the point where it is willing to accept the hominid record as indicative of the course of human evolution, it is the professional anthropologists who have tended to become uneasy at the possibility of discovering a Neanderthal skeleton in the *sapiens* closet. However, if we accept the Pithecanthropines as being a stage of human evolution, it is difficult to get from there to Modern form without going by way of something that must be regarded as Neanderthal. Add to this the occurrence of Neanderthals in some quantities in the time immediately prior to the earliest reliably dated appearance of humans of Modern form, then the probability that the Neanderthals were our ancestors is greatly increased.

Prehistoric research has been going on longer in western Europe than anywhere else, so it is no surprise to find that evidence for the evolutionary stage immediately prior to ourselves was first discovered and named in Europe, and that more Neanderthal remains have been discovered there than anywhere else—starting with the first recognized specimen in 1856, which gave its name to the whole stage. In spite of this, it has taken the better part of a century for the various areas of interest that constitute the science of prehistory to mature.

In the meantime, many discoveries have been made that could not be adequately treated because of the limitations of the times: stratigraphy was not controlled; faunal or cultural associations were not recorded; absolute dates could not be determined; and so on. Consequently, despite the quantity of Neanderthal from Europe, virtually none of the major specimens can be precisely placed in time. The best we can do is associate them with the Mousterian tool-making tradition, which, in turn, can be dated from approximately 35,000 years ago back to the beginning of the last glaciation at the end of the last interglacial some 100,000 years ago. However, direct dating of Neanderthal skeletal material has been made at two sites in the Middle East, one in Israel (Tabūn) and one in Iraq (Shanidar). On the basis of ^{14}C determinations, both can be assigned an age of 50,000 to 60,000 years ago. No fully developed Neanderthals are more recent than this, and the evidence suggests that many are much older, so it is fair to use this date as the tentative boundary for the most recent occurrence of the stage as a whole. For the beginning of the stage, specimens are few and fragmentary, and since they belong to a period too old to be dated by ^{14}C and too young to be dated by K/A, there is more than a little uncertainty remaining.

NEANDERTHAL FORM

To consider the form of the Neanderthals, we must start by dispensing with the image of the hairy, slouching beast, tramping through the Ice Age snow drifts clad only in a loincloth, and not quite able to stand erect. Although there are minor differences between the pelvis of the known Neanderthals and that of modern humans, there is no evidence to indicate that Neanderthals' posture was any less erect than ours. The human line has stood upright since the Australopithecine Stage, and any attempt to inflict the Neanderthals with a "bent-knee gait" is simply a survival of the efforts on the part of early interpreters to view all aspects of Neanderthal

A Neanderthal. The classic "cave man"—leopard skin, club, and all—dimly peering at a world that is largely beyond his comprehension. In fact, the Neanderthals probably had much more effective weapons and clothing, and there is reason to believe that they were at least as intelligent as modern humans, if not more so.

anatomy as being "primitive" or ape-like. (This also involved the incorrect assumption that apes cannot straighten their legs at the knee.) From the neck on down, the main difference between Neanderthal and Modern form is the indications of generally greater ruggedness in Neanderthal joints and muscles. As with the Pithecanthropines, the demands imposed on the Neanderthals by a Middle Pleistocene hunting way of life continued into the early part of the Late Pleistocene. This is reflected in the continued existence of a powerfully developed skeletomuscular system, particularly in the males. When we compare this to the modern situation, it is evident that this is more an average difference in degree rather than a difference in kind.

One postcranial detail has led to a recent round of speculation and discussion. This is the fact that, while most of the postcranial skeleton differs from Modern form simply in being more robust, the superior branch (ramus) of the pelvis seems curiously elongated. In the female pelvis, this has the effect of enlarging the birth canal, and it has led one otherwise respected anthropologist to the unlikely conclusion that pregnancy in Neanderthal women lasted a full year instead of the normal human nine months. The elongated superior pelvic ramus was necessary to allow passage for the abnormally large newborn produced by that supposedly prolonged pregnancy.

A recently completed study, however, has demonstrated that no such unlikely deviation from the normally expected human biological condition need be postulated to account for what can be observed. The Neanderthals were merely extraordinarily robust for their stature—something that we have, in fact, known now for over a century—and people who are robust for their stature develop from, and in turn produce, relatively large newborn infants. The superior ramus of the pelvis in modern human

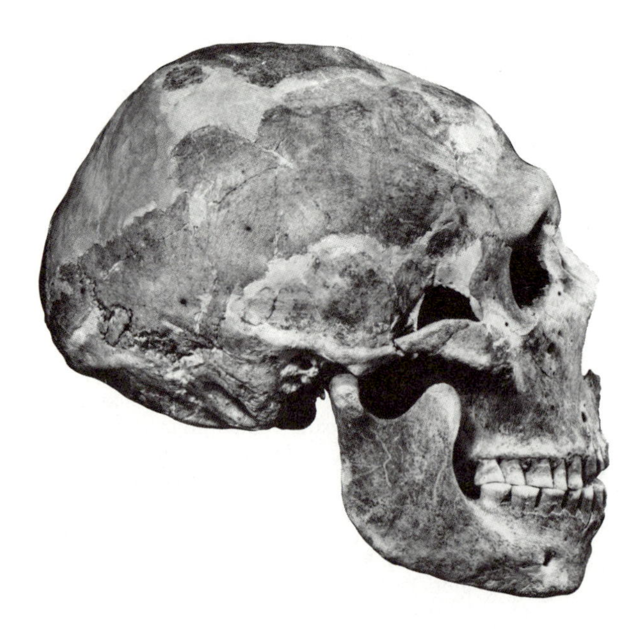

The skull of Shanidar I, a classic Neanderthal from Iraq. (Courtesy of the Iraq Museum, Baghdad.)

populations in whom the robustness to stature ratio is especially high, such as the Eskimo, deviate from the average modern human condition in just the direction that can be seen in its extreme manifestation in the Neanderthals.

Above the neck, however, the differences are far more marked and obvious. To be sure, the cranial measurements of some Neanderthals do not surpass those of some Moderns, but other Neanderthals present an array of dimensions that cannot be matched in recently living people. These revolve around the dentition and associated facial areas, where the Neanderthals do not differ functionally from the Pithecanthropines. All told, the Neanderthals are distinguished from the Pithecanthropines mainly by the possession of braincases of fully Modern size, while they are distinct from Moderns in the possession of Pithecanthropine faces and teeth. In fact, the Neanderthal front teeth include the largest to be found in the whole picture of human evolution, although this may simply be due to the scarcity of specimens from the earlier stages.

BRAIN SIZE AND THE ORIGIN OF LANGUAGE

Some people have regarded it as puzzling that the human brain should have attained full size 100,000 years ago or possibly more and remained the same ever since. The argument has been advanced that, if intelligence has survival value, more intelligence should have greater survival value. This, however, is failing to recognize that "survival of the fit" is a more appropriate expression than "survival of the fittest," and that the primary human adaptive mechanism is culture. When culture had developed to the point where the knowledge and traditions transmitted would confer an adequate chance of survival on any who could master it, the advantage of being yet more intelligent became relatively unimportant. Although we could argue that an innovator must have more intelligence than a person who is just able to master the culture in which he or she is brought up, it still remains true that the dullest member of a group benefits from the innovation of the brightest to an equal extent, and that genetic endowments are passed on to the next generation with the proportions unchanged. In the face of such an explanation, it would be surprising to find an increase in cranial capacity during the last 100,000 years.

This, then, suggests that brain size should have ceased enlarging as soon as an effective means of passing on cultural traditions had been developed. Of course, the most effective means of accomplishing this is through the phenomenon of language as we know it. A good case can be made that this coincides with the beginning of the Neanderthal Stage. Throughout the preceding Middle Pleistocene, brain size had been increasing at a slow but steady rate. Clearly the additional information-storage capacity which this represented was advantageous to the possessors. Language, however, added a completely new dimension that vastly transcended the advantages of raw storage capacity. With language, which acts in a sense the way a compiler program does for a computer, information can be processed and transferred, and comparisons and considerations can be made between all the communicating members of a group. The verbal world that we now all take for granted thus became a reality for the first time. If this required a particular biological threshold, I would argue that this had been achieved by the Neanderthals, and that the basis was now provided for all of the accelerating and accumulating cultural achievements that have subsequently taken place.

NEANDERTHALS AND THE MOUSTERIAN

But what is a Neanderthal? A tentative definition has been offered: "Neanderthals are the people associated with the Mousterian culture just before the reductions in form and size of the Middle Pleistocene faces and teeth." Middle Pleistocene levels of robustness and muscularity were still evident, but the brain was fully Modern in size and, presumably, function.

Because part of this definition is, in fact, cultural, we next have to take a closer look at what is under consideration. The term "Mousterian" comes from the village of Le Moustier in southwestern France where the type site is located. Tools of Mousterian form are distributed throughout western and southern Europe, south through the Balkans, east into the Middle East, and northeast through the Crimea, the Caucasus, and Uzbekistan. Throughout this whole area, which we could call a Mousterian culture area, there were a series of bands possessing related cultures between which similar cultural elements maintained circulation. Local differences in details of typology and technique of manufacture persisted, but all these subcultures possessed the same functional tool categories: scrapers, points, and knives.

Scrapers indicate a concern for the preparation of animal hides, which is reasonable for people living in a subarctic climate. It used to be thought that effective clothing was not developed until the ensuing Upper Paleolithic, with the invention and manufacture of bone needles, but there is no reason to deny the Neanderthals the use of skin clothing just because they had no needles; wrapped clothing bound on by thongs was utilized by the poorer peoples of Europe right up to historical times. Certainly the Neanderthals must have been doing something with the skins they went to so much trouble to prepare, and it is reasonable to suppose that the manufacture of clothing was one such thing. Indeed, human survival in Europe during the last glaciation without clothing would have been impossible.

The Mousterian points, made on flakes of a variety of sorts, evidently were frequently hafted, and the inference can be made that spears were being so tipped. Whether these were thrusting spears or throwing spears we have no way of knowing, but they obviously played an important role in the Neanderthal way of life. We might comment that the complex and precisely coordinated activities associated with throwing comprise a uniquely human phenomenon. No other animal has attained any degree of effectiveness in this practice at all. Possibly this is symbolized by the Mousterian point, but there is no way to prove this, however tempting it may be to add this to the evidence for increased hunting efficiency by the Neanderthals.

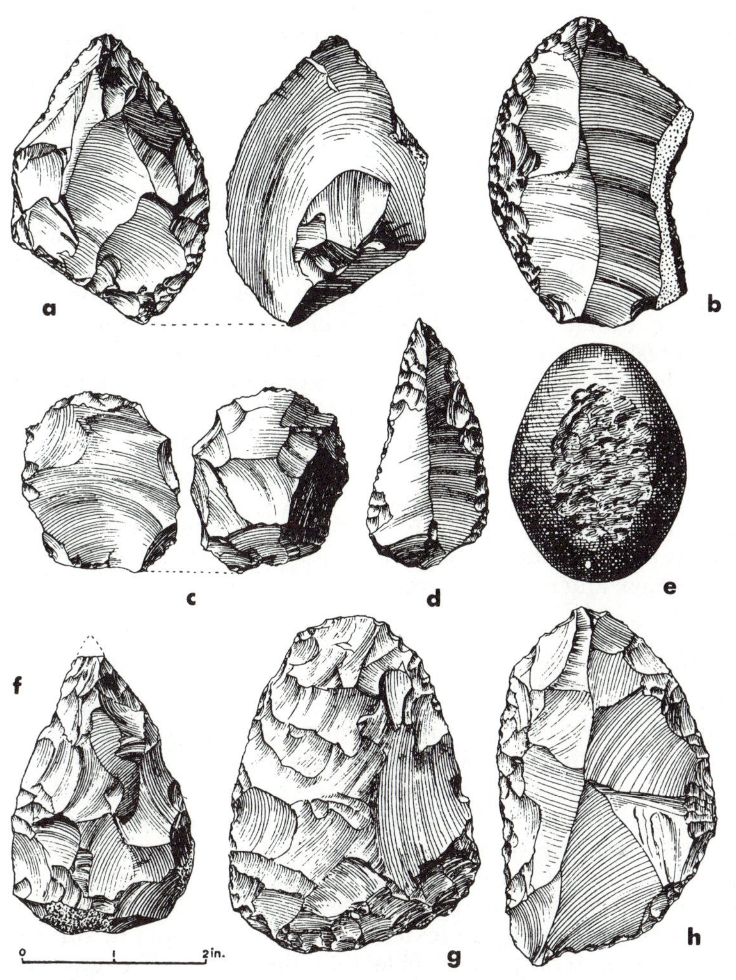

A collection of Mousterian tools: (a,b) side-scrapers (racloirs); (c) disc-core, and (d) point, from the rock-shelter at Le Moustier near Peyzac (Dordogne); (e) small anvil or hammer-stone (pebble of ferruginous grit), Gibraltar caves; (f) hand-axe from Le Moustier; (g) hand-axe (chert), and (h) oval flake-tool (flint), from Kent's Cavern, Torquay. (a–d) Typical Mousterian; (f) Mousterian of Acheulian tradition; (g, h) of Acheulo-Levalloisian tradition. (By courtesy of the Trustees of the British Museum of Natural History.)

TEETH AS TOOLS

The onset of the last glaciation must have increased the problem of simple survival for the inhabitants of the northern portions of the Old World. Chances of survival obviously would be improved as people increasingly attempted to manipulate and shape the natural world confronting them. Two factors indicate an increasing concern for manipulative behavior among the Neanderthals. One is the heavy rounded wear that appears on their front teeth. Not only that, the front teeth reach the maximum size that they achieve during the course of human evolution, evidently in response to the selective pressures that favored the development and maintenance of that anatomical manipulating device, the original built-in. Even the casual observer cannot fail to be impressed by the extensive wear often found on Neanderthal front teeth, which are frequently eroded to such an extent that the entire enamel crown has been worn down so that the surviving worn stumps are just what remains of the roots. Moreover, a detailed examination using a scanning electron microscope has revealed that the flaking and scratch marks on the enamel of worn tooth crowns is remarkably similar to that visible in recent Eskimos, who are famous for the manipulative tasks to which they put their teeth. The only difference is that the Neanderthal teeth display an even more extensive amount of chipping, flaking, and scratching on their enamel.

The survival value of manipulative behavior is easy enough to appreciate, and, eventually, the use of the teeth as all-purpose tools begins to give way to the employment of special tools for special purposes. This becomes much more obvious as the Mousterian gives way to the Upper Paleolithic with its proliferation of special tool categories—and a reduction in the extent of heavy anterior tooth wear.

SURVIVAL IN THE NORTH

Before the onset of the last (Würm) glaciation, the representatives of genus *Homo* were unable to cope with a subarctic environment. Consistent with their African area of origin, humans have remained physiologically tropical mammals to this day. The ability to invade and exploit other environments is a product of specializations in the cultural adaptive mechanism. Until late in the last interglacial period, however, this cultural adaptive mechanism was not well enough developed to compensate for human physical inadequacies to the extent of allowing people to survive in a really chilly area. It is just possible that the use of fire was not controlled to a sufficient extent that it could be reliably kindled and maintained to stave off the glacial chill. And the absence of quantities of scrapers suggests that the use of prepared hides as clothing against the cold had not yet been discovered. In any case, the onset of the preceding glaciations had forced people out of the increasingly inhospitable parts of what had formerly been the temperate zone. Climatic changes produced by the onset of glacial conditions were most extreme at the western end of the Old World temperate zone, where the Alps acted like an enormous refrigerator and cooled off the whole of Europe. Scandinavia added to this general chilling and contributed to the continental ice sheet, which moved south across the Baltic, blanketing the northern edge of continental Europe and much of the British Isles. Because the most extreme climatic changes in the Old World focused at the European end of the range, it is reasonable to expect that the greatest population dislocations occurred there as well.

CULTURAL DIFFERENTIATION

By the time of the onset of the Würm, however, the pre-Neanderthal level of cultural attainment was just high enough so that, with some modifications, it allowed people to remain in the more northern unglaciated parts of Europe, southern Russia, and the Middle East, and to take advantage of the abundant food supply represented by the great numbers of large Pleistocene mammals that thrived there. Culturally, this represents a kind of forced adaptation that took place in the western reaches of the north temperate zone, with the consequence that, for the first time since the Australopithecine Stage, there was a marked difference in the cultural adaptations of otherwise similar peoples in different parts of the world. This could very well indicate the start of the differential application of selective forces that produced the spectrum of variation which today is perceived under the rubric of "racial" differences.

Archaeological evidence from Africa suggests that some of the technological items of the Mousterian culture area were common throughout the regions to the south. This is also true in India and eastern Asia all the way to Japan, although in the latter instance it is not clear that those items of technology go back as far as the beginning of the last glaciation between 70,000 and 100,000 years ago. Haftable chipped stone points are then widely distributed throughout the Old World at the time when the Neanderthals were well established in the west, but the whole complex that bears the label Mousterian remained in the area from the Middle East through western Europe.

THE ORIGIN OF COOKING

Another cultural development at this time, also originally a part of the formal complex designated Mousterian, was eventually to change the face of humanity in a literal sense. This was the beginning of culinary elaboration. We are not really attuned to thinking of gastronomy as a Neanderthal invention, and perhaps it would be stretching things a bit to make such a claim, but there is reason to believe that the Neanderthals were the ones who pioneered the use of cooking as a regular means of preparing food.

Hearths appear in abundance in the Mousterian. Their frequency and extent clearly attest to their much greater importance, in contrast to the evidence from earlier habitation sites. They were also clearly being used for more than just keeping people warm. Further, the "hearths" that appear in Mousterian sites were obviously different in kind from those used for open camp fires. Even after 50,000 years or more, the swirl of ashes and fire-blackened cobbles of a Mousterian hearth are preserved to a depth which indicates that in original form it had been more than just a surface phenomenon. In fact, the remains look remarkably like those of recently used earth ovens (roasting pits) of the kind still relied on in modern Polynesia and elsewhere. Their construction and use follows this pattern: A pit is scooped out in the ground, a collection of fist-sized rocks is placed in the bottom, and a fire is built over the rocks. When the fire has consumed the fuel, the rocks are raked aside and the object to be cooked is placed in the bottom of the pit. The rocks are then pushed up against and over the food item, the whole is covered with a skin, or leaves, or grass (or burlap or canvas today), and dirt is shoveled over everything. The overlying dirt provides an insulating blanket that keeps the heat within, and, with the heat provided by the rocks, the food steams in its own juices. Aficionados of the New England clambake or the Polynesian luau claim that there is no tastier way of preparing food. The results are not only delicious, but they are also remarkably tender.

Most treatises on gastronomy and the culinary arts today stress the importance of the cooking process in producing the tastes that we value, but a good case can be made for viewing the relationship turned the other way around. We do indeed value the tastes produced by the cooking process, but the point of that process had nothing to do with the generation of a particular taste in the first place. The reason that things are cooked is to render them ingestible and digestible. Many vegetable foods cannot be handled by the human digestive system until they are chemically altered by the application of heat. Meat, on the other hand, does not have to be cooked at all to make it digestible. But after it has been sitting around for a while, particularly in warm weather and in the absence of refrigeration, it has to be cooked quite thoroughly or else it can cause considerable gastric distress, to say the very least.

Spoiled meat was probably not much of a problem for the Paleolithic hunters during their glacial winters, but frozen meat almost certainly was. Even the hungriest of Neanderthal bands could hardly have consumed more than a fraction of a woolly rhinoceros before the rest of it froze. Without some way of thawing it, the bulk of its meat would have been unusable, and no sapient Neanderthal would have undertaken the exertion, let alone the risk, of tangling with a Pleistocene rhinoceros merely for the sake of a single meal. The Neanderthals of the last glaciation obviously had to have used some sort of regular cooking techniques in order to make use of the majority of the meat acquired by their hunting efforts. The earth oven method is not only a logical candidate, but it is the one that makes the most efficient use of fuel. And then there are those Mousterian "hearths" that look remarkably like the remains of earth ovens. Surely this is more than meaningless coincidence.

COOKING AND DENTAL REDUCTION

The regular use of such a form of cookery would substantially reduce the amount of chewing necessary. This being the case, we could predict that the probable mutation effect would then be allowed to operate without detriment, and that dental reduction would shortly ensue. Again, this appears to have been the case. The largest collection of Neanderthal teeth is from the site of Krapina in Yugoslavia, which comes from the time just at or before the beginning of the onset of the last glaciation. The molars are fully as large as those of the Pithecanthropines a million years earlier, and the front teeth, reflecting the importance of their manipulative function, are even larger. But by the time of the Würm Neanderthals of

Belgium and France, maybe 40,000 years more recently, tooth reduction had proceeded to such an extent that, in gross size, they were closer to the average for Upper Paleolithic populations. It would appear that the reductions that served to produce Modern face form had already begun in the Neanderthal groups in the northern parts of their area of occupation, and it may very well have been the result of their innovative culinary practices.

As with the advance from the Australopithecine to the Pithecanthropine Stages, the development from the Pithecanthropine to the Neanderthal Stage took place throughout the inhabited parts of the Old World at the same time. This presents something of a contradiction if we use a strict interpretation of Mousterian. A better cultural term might be preferable for the purpose of defining Neanderthal as a worldwide stage. Some archaeologists have used the term Middle Paleolithic to differentiate it from the Upper and Lower Paleolithic, and this term might be preferable, were there not so much archaeological opposition. Perhaps the term "Mousterioid" might be used provisionally to include the Mousterian proper and all the similar cultures based on flake technology.

THE NEANDERTHAL DISTRIBUTION

As an indicator of the geographical distribution of the Neanderthal Stage, human skeletal material is almost better than the archaeological record—far less complete, of course, but more clearly indicative. The European

Neanderthal distribution as represented by the locations of the more important discoveries. 1. Neanderthal. 2. Spy. 3. La Chapelle-aux-Saints. 4. Le Moustier. 5. La Ferrassie. 6. La Quina. 7. Hortus. 8. Gibraltar. 9. Krapina. 10. Saccopastore. 11. Monte Circeo. 12. Mount Carmel. 13. Shanidar. 14. Teshik Tash. 15. Maba.

skeletal material is represented by the original Neanderthaler, the Spy remains, the "Old Man" of La Chapelle-aux-Saints, skeletal remains from La Ferrassie, Gibraltar, La Quina, Monte Circeo, and a great many other less complete finds. Relatively abundant remains have been discovered in southern Russia and the Middle East; perhaps the most exciting (and datable) remains come from Shanidar cave in Iraq. The most complete skeleton is the female from the Tabūn cave on Mount Carmel in Israel, which shows that sexual dimorphism was still pronounced, although perhaps just a little less so than it had been during the Middle Pleistocene. Skeletal remains of the Neanderthal Stage from the rest of the Old World are much less abundant, although the available fragments allow the inferences of distribution to be made. In China, for example, besides the Dali specimen that shows the Pithecanthropine-to-Neanderthal transition, there is a later full Neanderthal specimen from Maba in northern Guangdong (Canton) Province.

The foregoing should suffice to indicate the form, the dating, and the distribution of the Neanderthal Stage. Because full modern cranial capacity and more had been attained, presumably indicating intellectual capabilities at least the equivalent to those of modern humans, it would not be justifiable to regard the Neanderthals as specifically distinct from people of today. Formally, then, this makes them *Homo sapiens* with at most a subspecific appendage of *neanderthalensis.*

chapter fourteen

The Modern Stage

THE MYTH OF TRADITIONAL ANTHROPOLOGY

Fifty years ago, the appearance of modern humanity would have been presented in phrases such as these: "Sweeping into Europe from out of the East came a new type of man, tall and straight, with strong but finely formed limbs, whose superiority is proclaimed in the smooth brow and lofty forehead, and whose firm and prominent chin bespeaks a mentality in no way inferior to that of ourselves. In this fine and virile race we can recognize our own ancestors, who suddenly appear upon the scene and replace the degenerate and inferior Neanderthals, perhaps as a result of bloody conflict in which the superior mentality and physique of the newcomers tipped the balance. Whatever the cause, the lowly Neanderthals disappear forever, and the land henceforth becomes the never-to-be-relinquished home of our own lineage, the creator of the culture which is our own patrimony, and the originator of what has been built to the heights of Western civilization."

While this paragraph is pure invention, it nevertheless captures some of the flavor of the interpretive accounts of human evolution written a half century ago. Their appeal to the imagination of the literate world was immense. In the first place, the reference to an Eastern origin strikes a powerful chord in the mind of the Western reader, who is conditioned from infancy to regard all that is civilized and sanctified in antiquity to have had its origins "in the East." To these holy overtones are added the implications of the mysterious Orient. But this is just the beginning. The appeal to the lofty brow, often accompanied by explicit statements concerning the degree of development of the frontal lobes of the brain, caters to a folk belief, dating from the phrenology of the early nineteenth century and still current, that this is somehow indicative of superior mental ability. The portrayal of our own ancestors in terms which correspond to the stereotyped picture of European masculinity—prominent chin, tall, straight-limbed, to which hints of blue eyes and fair hair are often added—stimulates a conscious pride in belonging to such a line, a view that contains not a little racism as well. To complete the scene with all the components of a good old-fashioned melodrama, the Neanderthals are brought in as the embodiment of the villain. They are depicted as strong and dangerous, although dwarfed and physically "inferior," crafty but not really intelligent, hairy and doubtless violent and bestial. In spite of adversities, good prevails and evil is vanquished, and the Neanderthals disappear forever. As an added attraction to this already potent little drama, all direct relationship between the Neanderthals and the invading Moderns is either flatly denied or pushed so far back in time that it is lost in the mists of remote antiquity, which, of course, means that even people who are uneasy about accepting an evolutionary account of modern human origins can accept this story without any qualms.

Since I outlined this scenario in the first edition of this book, using a mixture of the prejudice and prolixity that sometimes characterized the literary style of a couple of generations gone by, two authors have picked up on it and rendered my little Paleolithic melodrama in modern dress—or, rather, undress, since that sells a lot more books, and a pot-boiler laced with a lot of explicit and steamy sex will find a far bigger market than elegantly phrased musings on the shape of brows and chins, even though both approaches may be expressing the same assumptions. Jean Auel's *Clan of the Cave Bear*, for example, recounts in graphic detail the brutal lust of the dim-witted Neanderthal who has his way with the "Modern" heroine. And in a much more literate version of Neanderthal/Modern encounters, Björn Kurtén depicts the lust of a female Neanderthal for the "Modern" object of her affections in his *Dance of the Tiger*. In both instances, however, the poor, dim Neanderthals have not quite gotten the hang of fully articulate speech, their technology is backward, and their

An Upper Paleolithic hunter, the first of the Modern Stage. Neatly dressed and clean shaven, he strides forth confidently to fulfill the destiny which his clear vision tells him is to be his future. Actually, the archaeological evidence does provide support for the existence, if not the invention, of tailored clothing and compound weapons in the Upper Paleolithic, but the lofty brow and "noble" expression are quite unwarranted idealizations.

impending doom is near. Even the hybrid offspring that result from the torrid scenes that sell the books are at best sterile and cannot perpetuate themselves.

Even though the style of popular presentation has gone through a quantum change in the past two generations, underneath it all there is little, if any, change in the basic message being conveyed, and the public is just as happy with it now as it was then. Dazzled by such dramatics, few people have been disturbed by the total lack of any reason for such a picture for invasion, of any source for the invaders, or of any perspectives on what they evolved out of and why. Analogous to the legend in which Athena sprang fully armed from the brow of Zeus, it would seem that twentieth-century prehistorians have tried to solve their headache concerning human origins by projecting a Modern human stereotype, fully formed from their own inner consciousness, smack into the early Pleistocene—thereby creating their own anthropological mythology. And even in the recent work specifically entitled *The Myths of Human Evolution* by Niles Eldredge and Ian Tattersall, the "myth" referred to in the title is not the scenario I have just presented—which they accept as demonstrated truth—but the possibility that modern human form could have evolved by natural means from its immediate predecessors.

THE NEANDERTHAL-TO-MODERN TRANSITION

It scarcely needs to be said that this book does not subscribe to this redefining of mythology. The attitude behind this presentation is based on the assumption that the hominid fossil record can be comfortably accommodated within the framework of standard evolutionary theory as it is applied to the human world. Noting that the Neanderthals have an antiquity demonstrably greater than that of Moderns, and that nothing but Modern skeletal material is evident since about 35,000 years ago, it is important to place both stages within the same evolutionary framework. If, as is claimed, the Neanderthals simply evolved into Modern form, then structurally and temporally intermediate forms should be apparent, and some rationale should be available to account for the change. Fortunately for the present scheme, both can be produced.

In the early 1930s, excavations in the rock shelter of Skhūl on the slopes of Mount Carmel in Israel (near the cave of Tabūn, which yielded a full-scale Neanderthal) produced a population of what can be called Neanderthaloids, because they recall genuine Neanderthals in many respects, but in other features, they deviate in the Modern direction. The dentition and the entire surrounding face has been somewhat reduced, leaving the forehead and sides of the cranial vault more vertical and producing the first vestiges of a distinct chin—formerly regarded as the "hallmark" of Modern form. Reductions in the robustness of ribs, long bones, and other aspects of the postcranial skeleton also show modification in the Modern direction.

For many years the suggestion was made that the Mount Carmel material belonged in the last interglacial, which would have made it older than the full Neanderthals dated to the Würm glaciation in Europe and recently confirmed at Shanidar in Iraq. To explain this mixture of traits, the interpretation was advanced that the people of Mount Carmel were hybrids between a fully Neanderthal group, represented by Tabūn and now Shanidar, and a fully Modern group, for which such vague fragments as Piltdown were advanced as documentation. Within the last twenty years, however, the Mount Carmel caves, especially the long stratified sequence at Tabūn, have been restudied using modern techniques and the dating has been revised. Using ^{14}C, the Tabūn skeleton has been shown to be around 60,000 years old, and the Skhūl material has been placed at about 35,000 years ago. This makes the Skhūl population intermediate in time as well as in form between the Neanderthal and Modern ends of the spectrum and eliminates all need for hybridization theories with their attendant difficulties.

More recently, excavations at Vindija in Yugoslavia have produced a quantity of fragmentary but interpretable remains that are contemporary with the Skhūl Neanderthaloids of Mount Carmel. These also show the same degree of archaic and modern traits and confirm the fact that a transition from Neanderthal to Modern form was going on in southeast Europe simultaneously with the transition taking place in the Middle East. And in western Europe, extensive if fragmentary remains of the latest Neanderthals from Hortus, in southern France, show clear-cut reductions in dental metrics and other aspects of morphology. At 35,000 to 39,000 years ago, they stand between full "classic" Neanderthal and early "Modern" levels of robustness.

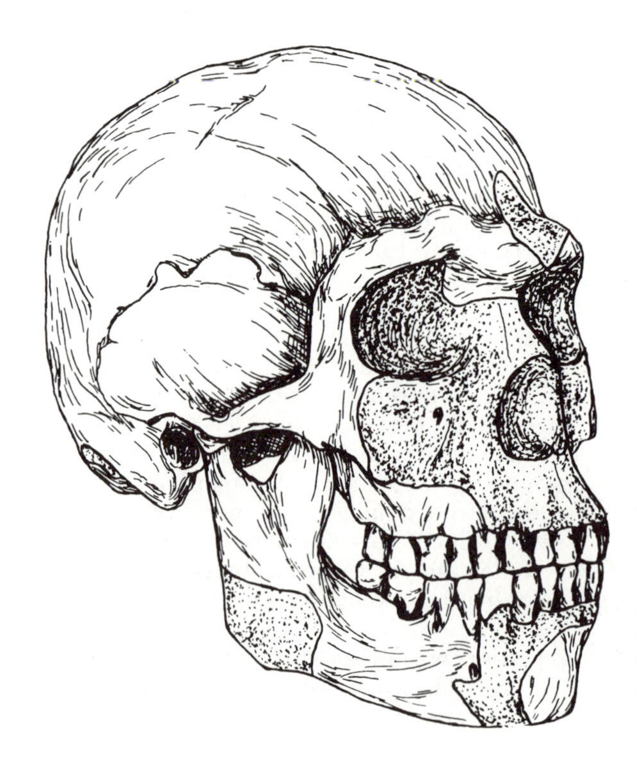

A Neanderthaloid skull, Skhūl V, the best-preserved representative from a group of ten individuals found in a rock shelter on the slopes of Mount Carmel, Israel, in the early 1930s.

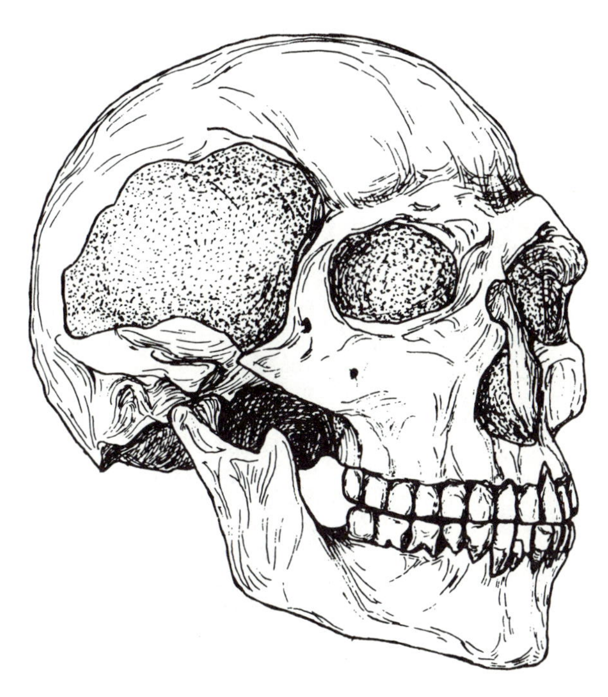

An Upper Paleolithic skull, Předmost III, a male from a large collection excavated in western Czechoslovakia in the late nineteenth and early twentieth centuries. A lingering robustness of brow ridges, facial skeleton, and muscle markings recall earlier conditions in human evolution.

A few other isolated finds of intermediate character also exist, and the Neanderthal origin of modern humans is further supported by the presence of a quantity of dissociated Neanderthaloid characteristics in the earliest clearly Upper Paleolithic populations. The first such Upper Paleolithic population to be discovered was the Cro-Magnon group, found less than a decade after Darwin's *Origin* appeared. Occasionally the term Cro-Magnon is applied to designate the early Moderns as a stage, and occasionally also it and other terms (Grimaldi, Combe Capelle, Chancelade,

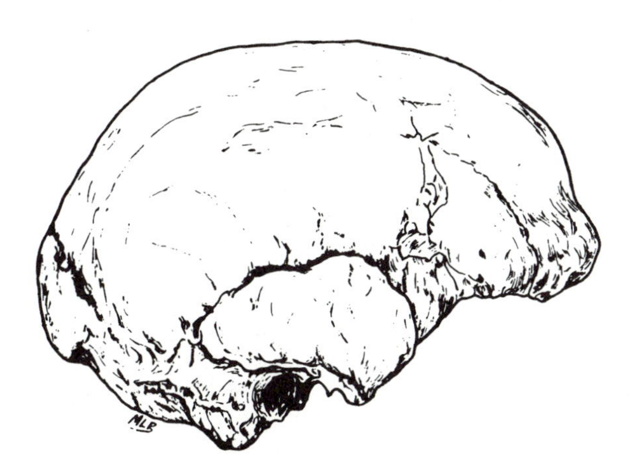

An early Upper Paleolithic skull from Mladeč in Czechoslovakia showing more than traces of Neanderthal form in the area of neck muscle attachments and in the brow ridges. The original was destroyed by German soldiers in World War II. Pictured is Mladeč V. (Drawn from a cast.)

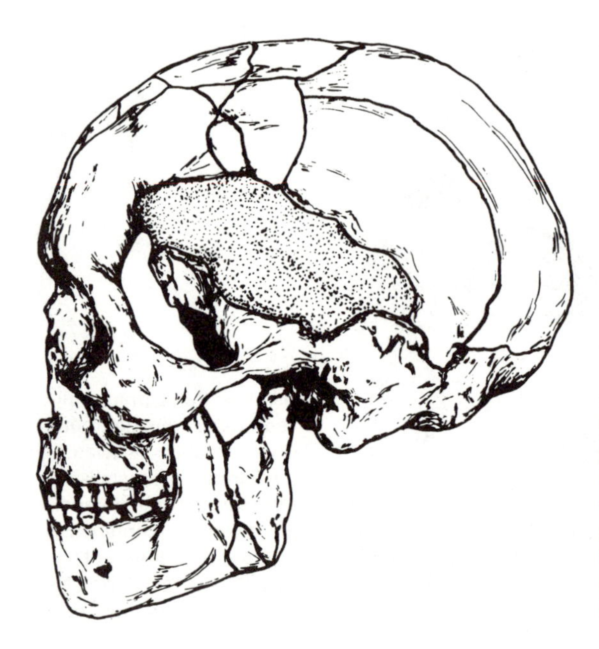

The Combe Capelle skull, an early Upper Paleolithic find from southern France showing some Neanderthaloid traces in the brows and in the tooth-bearing part of the jaws. The original was destroyed when an American bomb landed on the museum where it was stored in Berlin during World War II. (Drawn from a cast.)

etc.) are used to designate separate supposed "races" of Modern people in the Upper Paleolithic. This would seem premature, because it is most unlikely that four or more separate "races" existed in southern France (where these representatives were found) during the Upper Paleolithic. For the present, these can all be referred to as Upper Paleolithic representatives of the Modern Stage. As has already been mentioned, the chief physical differences between the Neanderthals and the Moderns is in the relative size and form of their dentition, its supporting facial architecture and related parts of the skull, plus certain aspects of general skeletal development and musculature. In all these features, the Moderns show a marked degree of reduction from the Neanderthal state, although the early Upper Paleolithic representatives are decidedly more robust in all of these features than are average Moderns today. To this extent, calling Upper Paleolithic form "Modern" obscures the fact that it is really intermediate between Neanderthal and twentieth-century degrees of robustness.

THE UPPER PALEOLITHIC AND ITS CONSEQUENCES

Several generations of scholars now have noted that the appearance of Modern form is correlated with the appearance of Upper Paleolithic tool-making traditions, although the modernity of that form is generally overemphasized. The advance in technological complexity of the Upper Paleolithic over the Mousterian is comparable to the advance the Mousterian showed over the Lower Paleolithic. Refinements in tool making were signalled by the technique of preparing flint cores so that long narrow spalls, technically called blades, could be detached. This increased the number of tools that a given amount of raw material could yield. Furthermore, the tools thus produced were worked into a greater variety of functionally distinct forms than was previously the case. Points, knives, and scrapers were refined, and to these were added a variety of gouging tools called burins. Evidently, these were used in working bone and antler, and they coincided with the appearance of an extensive bone industry: harpoon points, awls, and needles with eyes in them.

From the appearance of those needles, we can infer that shaped and sewn—tailored—clothing was being made. And with the evidence of the spear points and harpoon heads, it is apparent that hunters now definitely used the technique of hurling projectiles at prey. This is further supported by the appearance of spear throwers (*atlatls,* to give them their Mexican name), which, by acting as an extension of the arm, significantly increased the power of propulsion and added to the effective range over which a spear could operate. Certainly the vast quantities of animal remains found in Upper Paleolithic sites attest to the effectiveness of hunting techniques, and we can assume that a higher level of social cooperation lay behind the hunters' evident ability to drive game in quantity.

STRING AND WHAT IT IMPLIES

Not only is it evident that the people of the Upper Paleolithic were able to hunt large game with greater efficiency, but, for the first time in prehistory, small game began to appear in great and increasing quantities. Vast amounts of rabbit-sized mammals, birds, and fish were now being utilized, as attested to by the quantity of their bones in Upper Paleolithic sites. Of course, it is possible to spear a fish, a bird, or a rabbit individually, but this is a laborious and time-consuming process that does not yield the quantities to which the archaeological record now attests. What is indicated is the development of one of those things that we now take for granted but which has been called an "unobvious" element of technology—string. The concept of string is indicated by the presence of eyed needles. But string can also be used to make fish lines, noose snares, and nets. And with nets, flocks of birds and schools of fish become available sources of food to an extent previously impossible. Because birds, fish, and small mammals represent an edible biomass that is as great as or greater than that represented by large game, the subsistence base in the Upper Paleolithic was tremendously expanded, and at far less risk than had previously been the case. Casting a net over a flock of feeding quail or drawing a seine around a school of fish requires only a fraction of the raw muscle required to impale even a hundred or two hundred pounds of deer or wild pig and then survive the thrashings of the wounded prey.

With the advent of net-assisted hunting, the selective pressures that had long maintained Middle Pleistocene levels of skeletomuscular robustness were now markedly reduced. We could predict that the probable mutation effect would produce an immediate consequent reduction in that degree of robustness. This, in fact, is precisely what happened, and the predictable result is "Modern" form.

MOUSTERIAN CONTINUITY

Initially, the Upper Paleolithic appears in the same area where the Mousterian had flourished before it. The same caves were utilized as shelters, and the same kinds of animals continued to be hunted, although, as we have just noted, increasing quantities of smaller game were now being taken. Some Upper Paleolithic traditions in western Europe—the Châtelperronean, for example—are simply modified continuations of the local Mousterian. All told, the Upper Paleolithic can be regarded as a refined outgrowth—a culminating perfection—of the cold-climate adaptation of which the Mousterian represents the beginning. New technological items have been added, although many are simply refinements of the cruder Mousterian counterparts. That complex of refined cold-climate technological items and the whole hunting life way associated with them spread across the northern reaches of the entire Old World. After the end of the Pleistocene, they spread east across the Bering Sea to North

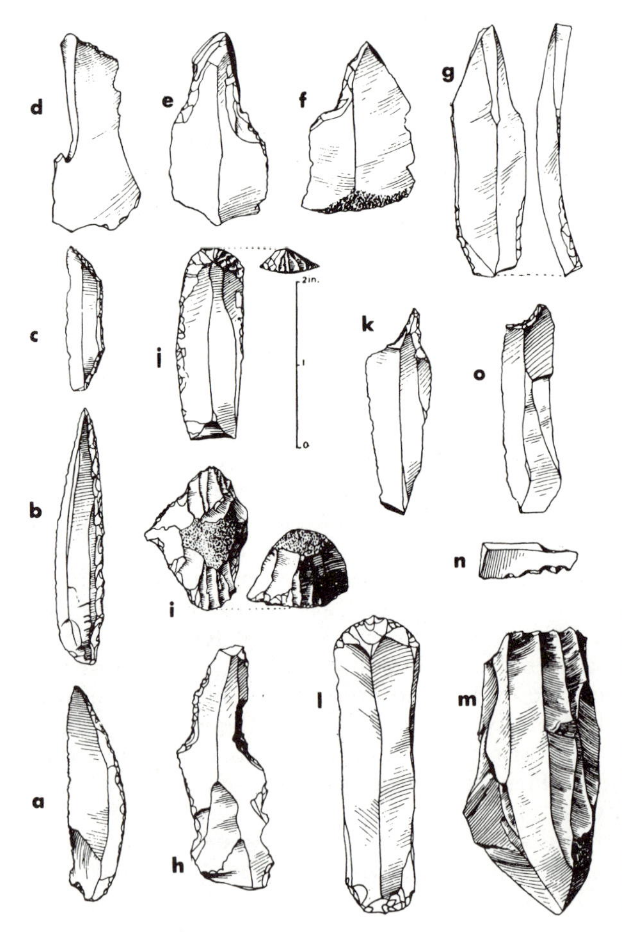

Upper Paleolithic flint tools: (a and b) "knife points"; (d, e, f, and g) gravers or burins; (i, j, l, and o) scrapers; (k) a piercer; (c, h, and n) miscellaneous tools; (m) a core from which blades have been struck. (By permission of the Trustees of the British Museum of Natural History.)

America, and their culmination could be seen in the extraordinary arctic hunting culture of the Eskimos from Alaska to Greenland, which lasted right up into the twentieth century, vestiges of which survive even today. Obviously, this was a remarkably successful cultural adaptation.

Survival in the north temperate zone during the Late Pleistocene depended upon the cultural developments that started in the Mousterian and continued without break in the Upper Paleolithic. Because cultural changes, even at the Australopithecine level, represent alterations in the selective forces that affect the hominids involved, we should expect to find some sort of reflected change to have occurred in the anatomy of the beneficiaries. I have already mentioned the consequences of the development of string technology. Another of the most obvious differences that set apart the Mousterian and the Upper Paleolithic from the preceding Lower Paleolithic is the appearance of a profusion of special cutting tools. From the choppers and bifaces (hand axes) and the crude flakes of the pre-Würm cultures, one goes to a variety of points, scrapers, and knives of the Mousterian to the even more elaborate stone technology of the Upper Paleolithic. Obviously, the ability to manipulate the surrounding world has been one of the prime factors in the successful survival of the human line, but equally obvious is the fact that a technological basis for any extensive manipulating did not exist prior to the Mousterian.

DENTO-FACIAL REDUCTION

From the extraordinary wear visible on the front teeth of the Neanderthals and their predecessors, we can suggest that the dentition bore the brunt of the finer manipulations and itself served as a sort of general all-purpose tool—the original built-in we have previously noted. With the appearance of an adequate cutlery at the beginning of the Mousterian, the significance of possessing large and powerful front teeth was substantially decreased. This would have allowed the probable mutation effect to operate throughout the early Würm, resulting in the reduction of the forward part of the dental arch and the supporting parts of the face. Furthermore, the process must have been speeded up as technological refinement advanced toward the Upper Paleolithic level. In similar fashion, the continuation of the cookery traditions that began in the Mousterian relaxed the selective pressures that had formerly maintained a Middle Pleistocene sized set of molars. The modern face, then, took shape as a result of reductions at both the front and back of the dental arch.

The significance of this change in tooth use can be appreciated if we consider for a moment the characteristic mode of eating of modern hunting and gathering people. Meat is not daintily manicured into bite-sized portions with knife and fork before ingesting. Rather, a chunk is taken in one hand, thrust part way into the mouth, where it is held with the front teeth, and then sawed off at lip level by means of a cutting implement: the "stuff-and-cut" school of etiquette, as it has been called. As practiced by modern hunters and a variety of peasants throughout the world, this procedure is aided by the efficiency represented by metal knives, but even so, the effect is sufficient to produce a substantial amount of flat wear on the incisors and canines, resulting in the "edge-to-edge" bite characteristic of so many of the non-Western peoples of the world. In fact, before the refinements of eating habits indicated by the adoption of chopsticks in Asia, and later by the knife and fork in the West, the edge-to-edge bite was the universal human condition. Before the development of metallurgy, this form of tooth wear was even more extensive; we can just imagine the burden placed on the front teeth *before* the development of even adequate *stone* cutlery.

The heavy wear apparent on the front teeth of many Neanderthals indicates that the burden was only gradually shifted from teeth to tools. For this reason, the reduction of the full Neanderthal face was only halfway accomplished in the Skhūl population at Mount Carmel and was still incomplete in the early Upper Paleolithic, which suggests that the teeth were still important for more than simply processing food. The curious rounding wear of Neanderthal incisors and the details of the manifestation of that wear revealed by the scanning electron microscope suggest that Neanderthals were using their teeth to process leather in a fashion similar to that of the modern Eskimo. This is consistent with the view presented previously that the Neanderthals were utilizing skins for clothing. This brings up another area of cultural adaptation and leads to another suggestive, if unproven, speculation.

THE START OF DEPIGMENTATION

Recall that the development of a hairless and heavily pigmented skin was suggested for the hunters early in the Pithecanthropine Stage. From this, we must assume that the early Neanderthals who first successfully adapted to the north temperate zone in the early Würm were dark brown or "black," as a correlate with the general human tropical physiology. The use of clothing, among other things, was of great importance in the success of their adaptation in the northern areas of habitation. By covering their skin with clothes, they drastically reduced the importance of the epidermal pigment melanin as an ultraviolet filter. Once again, the probable mutation effect operating over a substantial period of time would serve to reduce the structure whose importance had been decreased. In this case the "structure" is melanin, and its reduction resulted in depigmentation. It is of more than passing interest to note that, in general, those parts of the world where the amount of pigment in the human skin is at a minimum are also just those areas where Mousterian scrapers and Neanderthal teeth indicate that clothing has been utilized for the longest period of time. The picture evoked by a blond Neanderthaler is somewhat contrary to the usual stereotype, but it is quite possible that the invention of clothing by the Neanderthals of the early Würm was the source of the depigmentation phenomenon which allows some of the peoples of the world today to be described by the euphemism "white."

GRINDING STONES, POTS, AND CONTINUING EVOLUTION

Many have assumed that once Modern form was achieved, we had arrived, evolution had stopped, and we should have lived happily ever after, virtually unchanged, world without end, amen. The truth, however, is quite otherwise. The changes that produced early Modern out of Neanderthal form have continued in the same direction and, if anything, have actually speeded up. In fact, some aspects of dental and facial reduction are now proceeding at between double and triple the rate that served to convert Neanderthal into Modern form.

Late in the Pleistocene, the herds of large-sized game animals began to disappear. Some have suggested that the efficiency of the Upper Paleolithic hunters had gotten so great that they were cropping the herds faster than they could reproduce. In any case, big-game hunting became a less and less important aspect of human subsistence activities. Many an archaeologist, relishing the dramatic imagery conjured up by the Pleistocene mammoth hunters, has felt that the culture of the succeeding Mesolithic represented a "decline" from the heights of art and technology that had previously been achieved. Again, this simply is not the case. During the Mesolithic, the further concentration on nets and traps emphasized the shift in focus from the disappearing large game. Furthermore, the development of mortars and pestles and various forms of grinding technology allowed the people of the Mesolithic to begin to utilize an even more significant source of foods than had been possible before—namely, grains. The seeds of the grass family—wheat, oats, millet, barley, rice, and the like—as well as pulses such as lentils and peas, represent an enormous dietary resource that had previously been unavailable. Their exploitation led to a substantial population increase.

By the end of the Pleistocene 10,000 years ago, the use of grains as a subsistence base had advanced to the extent that they were being deliberately planted and tended in the Middle East and in South China/Southeast Asia, and the gathering of the Mesolithic had been transformed to the full-fledged farming of the Neolithic. Shortly thereafter, the discovery and manufacture of pottery further reduced the already lessened survival value of a large and well-formed dentition. In cooking pots, food can be simmered to drinkable consistency, at which point the edentulous can ingest their calories as easily as those who still possess a veritable Pithecanthropine dental arch. It is no accident that the populations with the smallest and fewest teeth in the world today are those associated with the areas where the Neolithic and preceding Mesolithic go back farthest in time.

With the suggestion that changes in the cultural adaptive mechanism are responsible for changes in face form and skin color, it should be possible to account for some of the major visible differences between the living peoples of the world in the same way. This is indeed the case, although to go into this in any detail is beyond the scope of this book. In the area where technological complexity has been having its impact for the greatest length of time, we would expect dental reduction to have proceeded to its greatest extent—and, as can be seen, these expectations are fulfilled. The most striking example is Australia, where the facial form of the Aborigines shows less reduction than it does in any other modern human group. Even within Australia, however, there is a remarkable north-south gradient. Mesolithic culture elements such as nets and seed grinders clearly came in from the north after the end of the Pleistocene. These altered conditions for the inhabitants, who had been living there for at least 40,000 years, to such an extent that, at the time of contact with the incoming Europeans, tooth size of the northernmost Aborigines had reduced to the level visible in the European Upper Paleolithic. That reduction was less apparent toward the south, where the incoming cultural elements had much less time depth. By the time one gets down to the Murray River basin in southern Australia, we find that tooth size had reduced only slightly from the Middle Pleistocene level of the fossil Murray inhabitants of only 10,000 years ago, and, at least in their molars, tooth size was distinctly larger than it had been for the "classic" Neanderthals of western Europe 50,000 years back.

Differences in the modern range of dento-facial robustness can be seen in the accompanying illustrations of male and female European

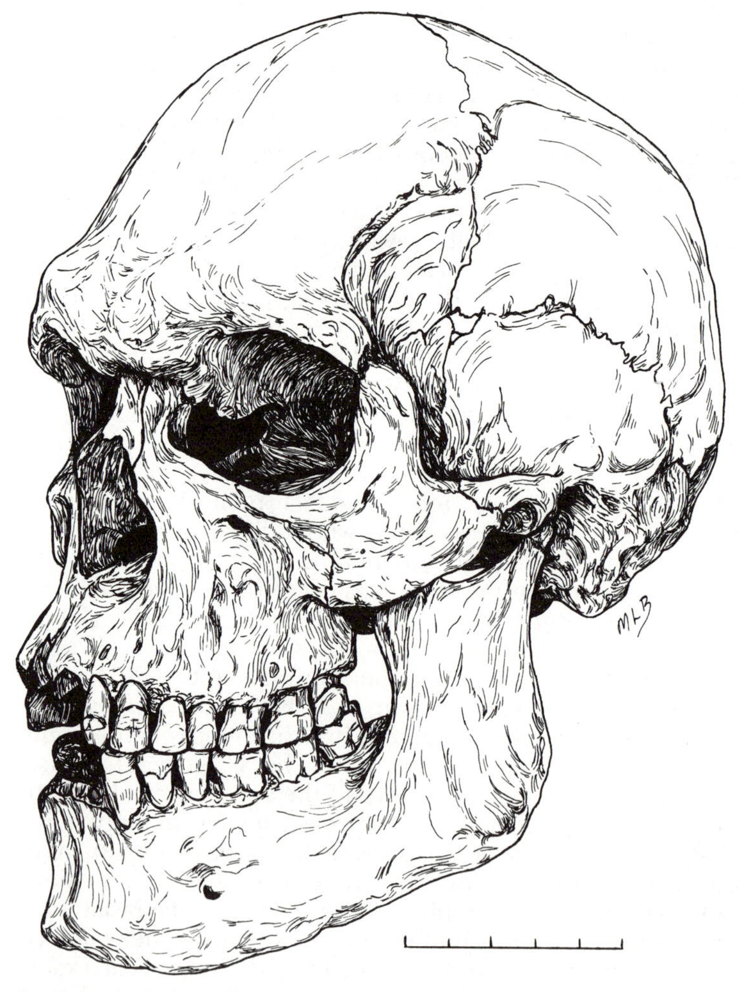

Southern Australian male. (Drawn by M. L. Brace at the Department of Anatomy, University of Edinburgh Medical School, courtesy of Professor George J. Romanes.)

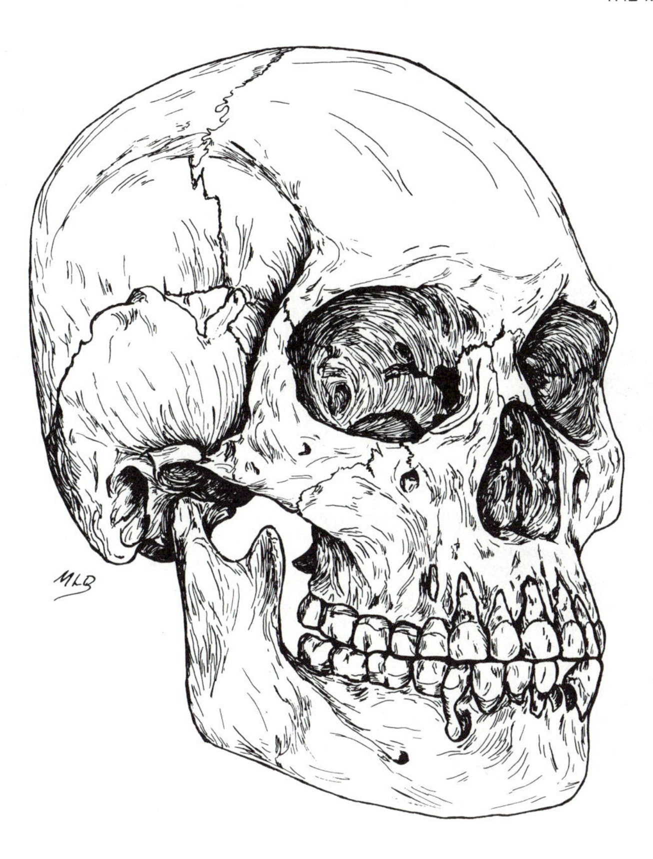

Southern Australian female. (Drawn by M. L. Brace at the American Museum of Natural History, New York, courtesy of Dr. Ian Tattersall.)

skulls—in this case, Germans—and the contrasting male and female Australian Aborigines—in fact, southern Australians from the Murray River basin. Admittedly, I chose one of the most robust Australian males for the example, so the brow ridge shown actually represents the extreme of the Australian range of variation. Even so, the degree of sexual dimorphism among the southern Australian Aborigines is considerably greater than that normally seen among the other modern human populations of the world. In this respect also, they have retained more of the characteristics generally found some 50,000 years ago.

The Europeans, on the other hand, do not represent the maximum degree of facial reduction among living human populations because, as can be clearly seen, they retain quite substantial noses, which shows a greater degree of retention from the past than is true for many of the other peoples of the world. The contrast between southern Australian and European tooth size and robustness, however, does come close to illustrating the extremes visible in the modern spectrum. The southern

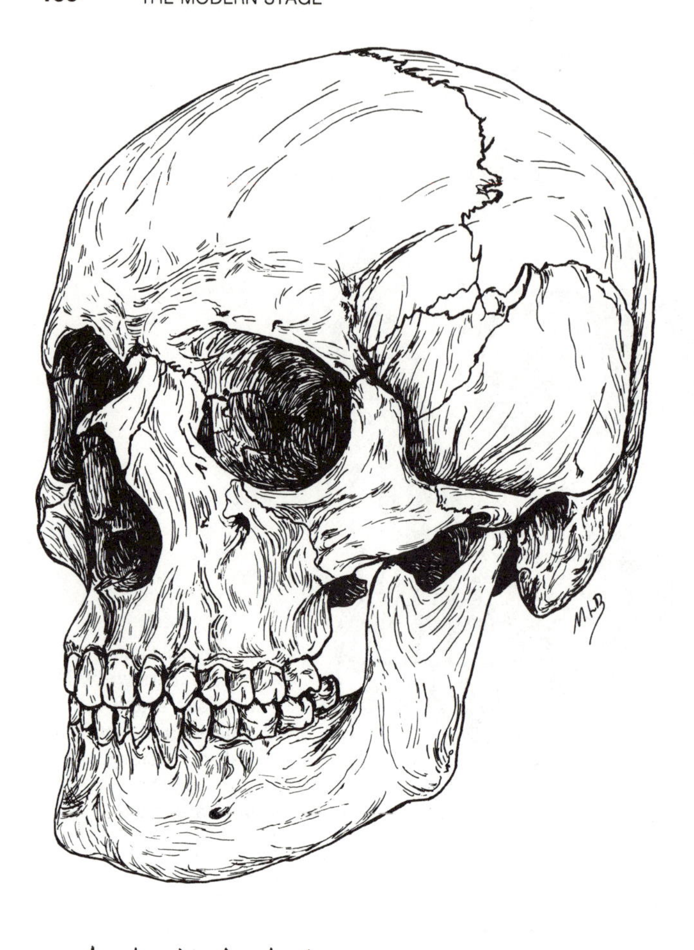

German male from Bavaria. (Drawn by M. L. Brace at the American Museum of Natural History, New York, courtesy of Dr. Ian Tattersall.)

Australians are clearly the least reduced in this respect, whereas the Europeans vie with just a few other groups as representing the most reduced. The illustration showing the contrast in the Australian and European condition displays the dental arches of the same female Australian and European individuals used to demonstrate the contrast with male cranial form depicted in the previous pictures. Not only are the teeth of the European markedly reduced in size, but the third molars (or so-called "wisdom" teeth) are congenitally missing, which reflects a kind of ultimate reduction. Australians, like all humans over 30,000 and more years ago, very rarely lack third molars. In modern Europeans and Chinese, as many as half the individuals examined will lack one or more third molars.

Using the arguments just developed, we can suggest that the human diversity visible in the world today is largely a product of events that have occurred during the last 75,000 years or so. It is only during this time that the archaeological record yields clear signs of functional differentiation in the cultural adaptive mechanism between one area of the world and another. As has been suggested, this was initiated by the survival problems posed by the periglacial areas. Solutions to these problems had a number of important consequences. One of these was the ability to thrive in the more northern areas that were opened up during retreats of the ice sheets. As they followed the melting ice sheets north, people at the Upper Paleolithic level of cultural development spread across the whole vast Eurasian steppe of the Old World and, at the eastern extremity, crossed the land bridge between Siberia and Alaska, producing the initial population of the New World—the only large-scale spread into previously unoccupied territory to have occurred since the expansion of the Pithe-

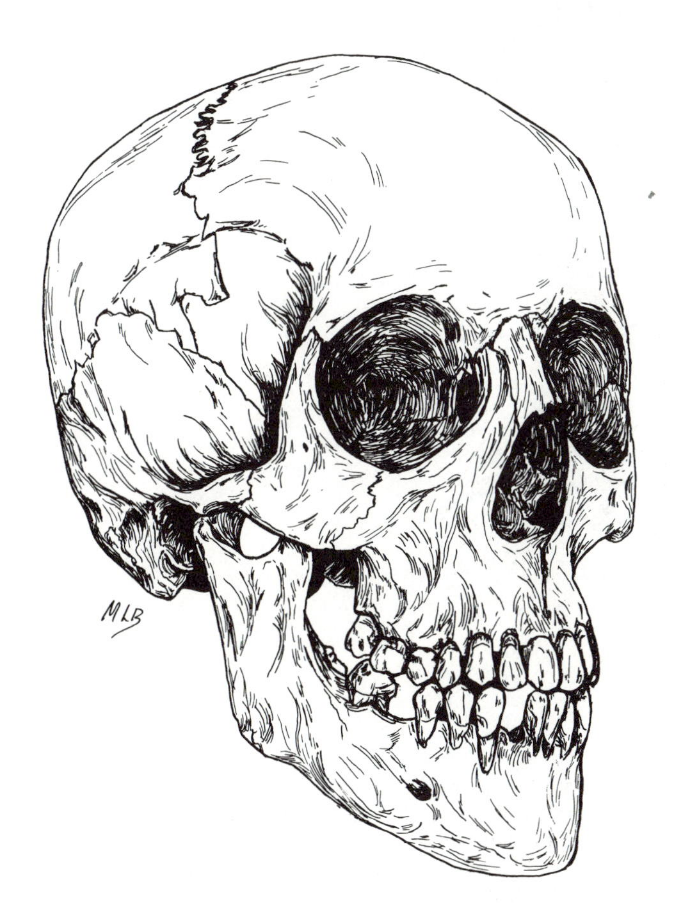

German female from Bavaria. (Drawn by M. L. Brace at the American Museum of Natural History, New York, courtesy of Dr. Ian Tattersall.)

canthropine stage throughout the Old World tropics. It was people at a developed level of this same general stage of complexity who were able to domesticate plants and animals, thus assuring their food sources and creating the foundation for the still greater cultural disparities that followed. (This food-producing or Neolithic revolution occurred about 10,000 years ago in the Middle East and Southeast Asia and, independently, somewhat more recently in Meso-America.) The effects of these various

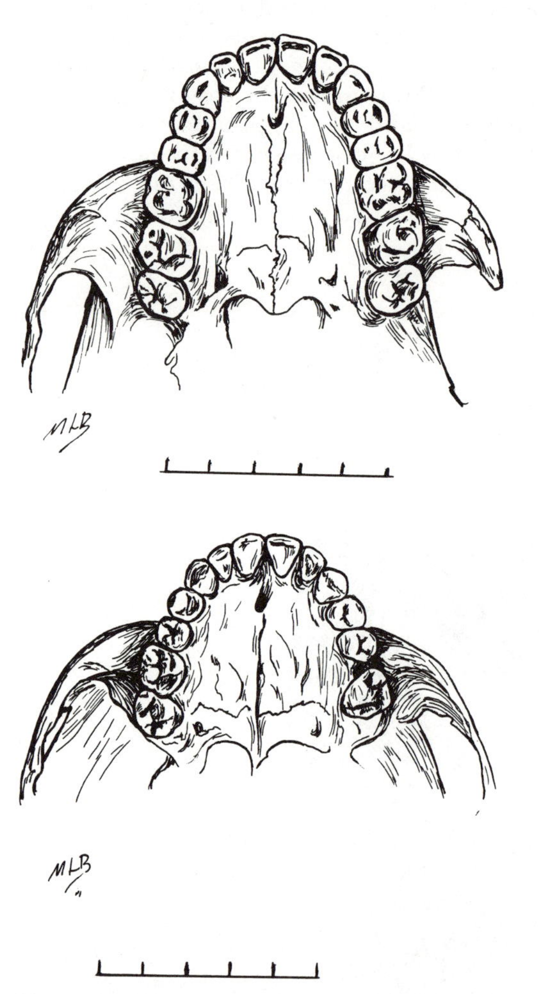

Size and robustness contrasts in Modern human palates and teeth. The large one is a southern Australian female, and the small one is a Bavarian German female. These are the same individuals that were used to represent the contrast in facial appearance in the previous illustrations. (Drawn by M. L. Brace at the American Museum of Natural History in New York, courtesy of Dr. Ian Tattersall.)

The man of the future (we are tempted to call him *Homo "durabilis,"* or the man who endures). Although it is perhaps unwarranted to visualize our distant descendants as puny, balding, myopic, and toothless little men, there is reason to suspect that some such trends may occur. Note that in the previous stages of human evolution the cultural element symbolizing human adaptation was held in the hand of the man in question, whereas in this portrayal it is hung around his neck.

developments slowly diffused into other parts of the world, but a detailed discussion of the events involved and their impact is the subject of other books by other authors.

We might be tempted to speculate that the increasing technical and medical ingenuity of developing world culture will further reduce or suspend the adaptive significance of many other human features. Reduction of these features as a result of the probable mutation effect would then follow, and it is possible to suggest that people in the future will be somewhat puny and underendowed by today's standards. Each era creates its own values, however, and the Neanderthals might very well have had the same feelings about ourselves, could they have known that we, their remote descendants, would be so much less robust than they. Perhaps our own hypothetical descendants of 1,000,000 A.D.—should we call them *Homo "durabilis,"* "the man who endures"?—will look back at the people of the twentieth century with feelings of repugnance and disgust. Tempting as such excursions may be, they do not properly belong in a book about the human past. In fact, they hardly belong to the realm of "science," and are included here only to lead the reader to realize that human evolution is not just something that occurred long ago: that it has been continual, that it is happening right now, and that it will go on in the future as long as people shall exist.

EPILOGUE

The figures on page 143 present a supersimplified picture of the changes occurring in the entire span of human evolution and their suggested causal correlates. The lower figure presents a record of hominid evolution seen from the point of view of tooth size alone. The vertical scale represents the average cross-sectional area in square millimeters of all of the teeth in the dental arch summed together. The robust Australopithecines had obviously become very large of tooth prior to their extinction. The spread at the *sapiens* end of the graph reflects the fact that within the last 100,000 years, the teeth of some human populations have undergone a marked degree of reduction, whereas those of others, the large-toothed southern Australian Aborigines, had scarcely been altered at all.

It does not take an expert to recognize that more than the usual amount of speculation has been included in this book. The major pieces of evidence have been presented, and evolutionary theory has been considered. The speculation entered when theory rather than solid evidence was used to support the interpretations offered, and it should be clearly recognized that this cannot constitute proof. As more fragments of human fossils are found in the years to come, the level of probability that one interpretation or another is correct will increase, but this is not proof either. Ultimately, it is impossible to "prove" the validity of any interpretation, because proof assumes an absolute certainty that can only occur in the realms of logic, mathematics, and religion, and not, in a formal sense, in science. At best, science can produce varying degrees of probability that its findings are to be believed. In interpreting the course of human evolution, the theoretical consistency of the scheme presented here should be justification enough for its development. The future alone can decide the probability of its rectitude.

Finally, for the cladists who feel that the hominid fossil record does *not* represent the course of hominid evolution, page 144 shows a cladogram that presents the four grades discussed in the preceding chapters as though they had branched from as-yet-undiscovered ancestors in the sequence shown. My own view, however, is represented by the final unilinear arrangement, where the Australopithecines evolved into the Pithecanthropines, which in turn evolved into the Neanderthals throughout the whole of the inhabited Old World, which finally became transformed into the various Modern populations alive today. I have left off the Australopithecine twig that became hyperrobust and died out before the Middle Pleistocene just to give a streamlined version of my general view.

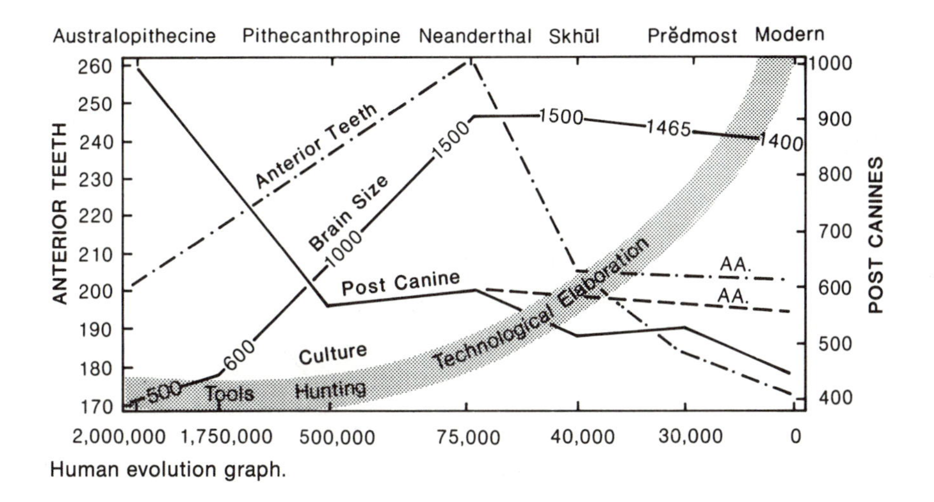

Human evolution graph.

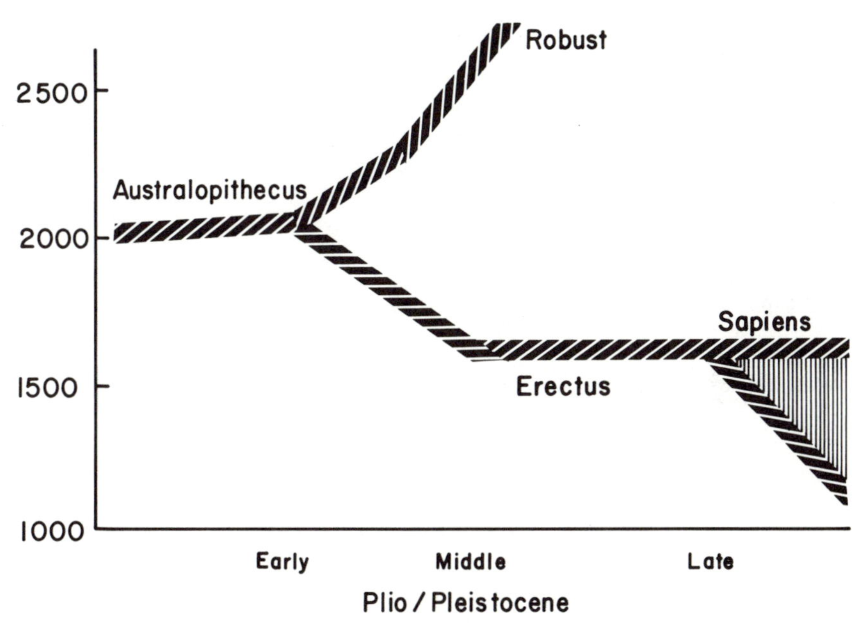

Hominid evolution plotted solely from the point of view of total tooth size. The vertical scale represents square millimeters of tooth cross-sectional area. The average cross-sectional area for each tooth category was summed to give the placement for each group named. The modern spectrum, which developed entirely within the last 100,000 years, shows the range of variation among the largest-toothed Australian Aboriginal groups and the groups with the smallest average tooth size, such as Chinese and Europeans. Although it shows how great the average differences are, these are small in comparison with the differences between the robust Australopithecines and the contemporary African Pithecanthropines. Each of the major changes on the graph correlates with major changes in adaptive strategy known for the groups indicated.

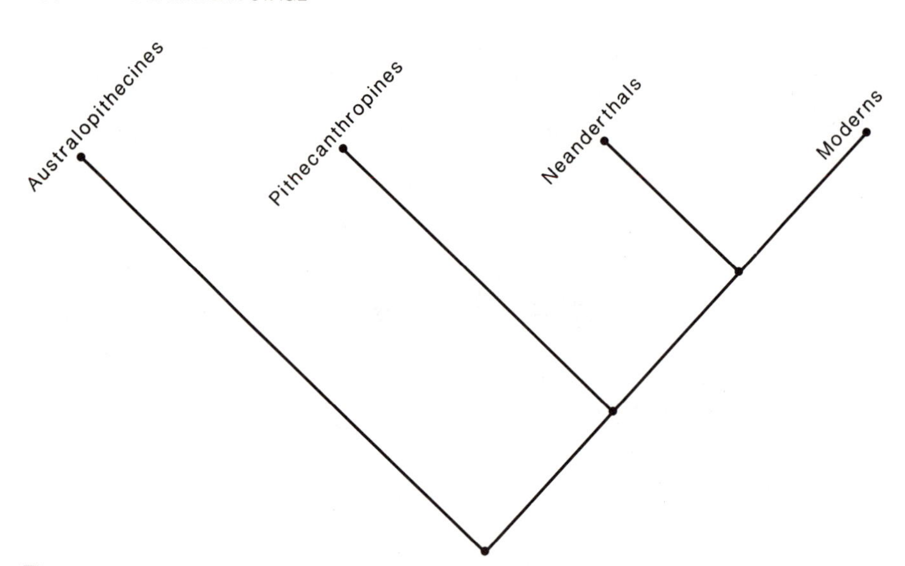

The stages of human evolution arranged in the form of a cladogram.

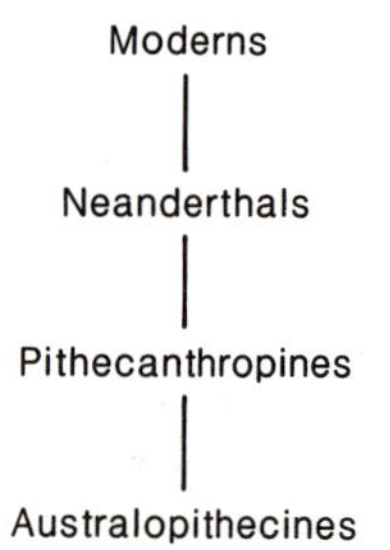

The stages of human evolution arranged in linear sequence.

Selected References

Chapter One. A few works will serve as background reading for evolution in general and human evolution in particular. The mind set that preceded Darwin's epoch-making publication is well displayed in that modern classic, *Genesis and Geology: A Study in the Relations of Scientific Thought, Natural Theology, and Social Opinion in Great Britain, 1790–1850,* by Charles Coulston Gillispie (Cambridge, Mass.: Harvard University Press, 1951). The reaction to Darwin's synthesis by the various intellectual communities in the world is examined in exemplary detail in the selections found in Thomas F. Glick, ed., *The Comparative Reception of Darwinism* (Austin: University of Texas Press, 1974). The most complete treatment of the background for the modern biological outlook is that massive *tour de force, The Growth of Biological Thought: Diversity, Evolution, and Inheritance,* by Ernst Mayr (Cambridge, Mass.: Harvard University Press, 1982). The resurgent and continuing fundamentalist Christian opposition to an evolutionary outlook is treated in excellent and critical fashion in Laurie R. Godfrey, ed., *Scientists Confront Creationism* (New York: W. W. Norton, 1983).

Chapter Two. The background for the development of evolutionary thought is well portrayed by John C. Greene in *The Death of Adam: Evolution and Its Impact on Western Thought* (Ames: The Iowa State University Press, 1959). The triumph of the Darwinian point of view is well developed in Loren C. Eiseley's *Darwin's Century: Evolution and the Men Who Discovered It* (Garden City: Doubleday & Company, Inc., 1958).

To gain some background in the attitudes of the French intellectual climate, where Darwinian evolution has been resisted right up to the present, see William Coleman, *Georges Cuvier, Zoologist* (Cambridge, Mass.: Harvard University Press, 1964). The continuation of such views is graphically demonstrated in several excellent essays, for example: "France" by Robert E. Stebbins in T. F. Glick, ed., *The Comparative Reception of Darwinism* (Austin: University of Texas Press, 1974:117–167); "Evolutionary Biology in France at the Time of the Evolutionary Synthesis," by Ernest Boesiger in E. Mayr and W. B. Provine, eds., *The Evolutionary Synthesis* (Cambridge, Mass.: Harvard University Press, 1980:309–321); and "A Second Glance at Evolutionary Biology in France," by Camille Limoges in E. Mayr and W. B. Provine, eds., *The Evolutionary Synthesis* (Cambridge, Mass.: Harvard University Press, 1980:322–328). These provide a valuable perspective for understanding the interpretations offered in one of the widely used background books dealing with the fossil evidence for human evolution, *Fossil Men,* by M. Boule and H. V. Vallois (New York: Dryden Press, 1957), which itself is just an update of the original work by Marcellin Boule, *Les Hommes Fossiles: Éléments de Paléontologie Humaine* (Paris: Masson, 1921).

Chapter Three. The first and still one of the best attempts to interpret the skeletal remains of prehistoric human populations from a systematic point of view is *Studien zur Vorgeschichte des Menschen,* by Gustav Schwalbe (Stuttgart: E. Scheizerbart, 1906). Probably the best account of early discoveries and interpretations in English is Aleš Hrdlička's *The Skeletal Remains of Early Man,* Smithsonian Miscellaneous Collections No. 83 (Washington, D.C.: The Smithsonian Institution, 1930).

Chapter Four. The first explicit attempt to consider the impact of national intellectual traditions and the accidents of history on the interpretation of the evidence for human evolution is in "The Fate of the 'Classic' Neanderthals: A Consideration of Hominid Catastrophism," by C. L. Brace in *Current Anthropology,* Vol. 5, No. 1 (1964). Particularly interesting are the irate comments of the proponents of the traditional view, which are printed following the main body of the article. This provides the basis for the view presented in Chapter 8 of C. L. Brace and Ashley Montagu's *Human Evolution: An Introduction to Biological Anthropology,* 2nd ed. (New York: Macmillan, 1977), and for the orientation of the present volume. The outrage of the orthodox to this stance is well displayed by the review of the first edition of Brace and Montagu (1965) by Robert W. Ehrich in *Human Biology,* Vol. 38, No. 3 (1966), and "More on the Fate of the 'Classic' Neanderthals," *Current Anthropology,* Vol. 7, No. 2 (1966).

Chapter Five. The accounts of those who contributed to the discoveries are particularly interesting. Franz Weidenreich reported some of his conclusions in *Apes, Giants and Man* (Chicago: The University of Chicago Press, 1946). A more complete personal account is presented by G. H. R. von Koenigswald in *Meeting Prehistoric Man* (London: Thames & Hudson, 1956), and perhaps the most charming personal story is told by Raymond Dart with Dennis Craig in *Adventure with the Missing Link* (New York: Harper & Brothers, 1959).

Chapter Six. The continuing contributions of Raymond Dart and the important discoveries that have come from South Africa are engagingly related by Dart's former student and successor as Professor of Anatomy at the University of the Witwatersrand Medical School, Phillip V. Tobias, in his *Dart, Taung and the Missing Link. An Essay on the Life and Work of Emeritus Professor Raymond Dart,* Based on a Tribute to Professor Dart on his 90th Birthday, delivered at the University of the Witwatersrand, Johannesburg, on 22 June 1983 (Johannesburg: Witwatersrand University Press, 1984), and "History of Physical Anthropology in Southern Africa," *Yearbook of Physical Anthropology,* Vol. 28 (1985).

Much of the progress made in East Africa is bound up with the work of the flamboyant Leakey family, and, fortunately, there are extensive biographical and autobiographical accounts. Sonia Cole has made a conscientious effort to chronicle the life of the most spectacular member of that extraordinary family, the late Louis Leakey, in her *Leakey's Luck: The Life of Louis Seymour Bazett Leakey* (New York: Harcourt, Brace, Jovanovich, 1975). Mary Leakey has written of her own remarkable and eventful life in a charmingly matter-of-fact fashion in *Disclosing the Past: An Autobiography* (London: Weidenfeld & Nicolson, 1984). Not to be outdone, Richard Leakey has also rendered a somewhat self-serving account of his own adventure-filled life in *One Life: An Autobiography* (Salem, New Hampshire: Salem House/ Merrimack Publishers Circle, 1984), although the impression of arrogance may leave the reader with a bit of an uneasy feeling.

The Leakey family may have dominated the East African scene, but they have not been without challengers. The spectacular finds in Ethiopia and the different interpretive slant that has accompanied them is recounted by Donald C. Johanson and Maitland A. Edey in *Lucy: The Beginning of Humankind* (New York: Simon and Schuster, 1981), although, as with the accounts of the Leakeys, the aura of personality and ego sometimes tends to take on a more central importance than the science that should be the justification for it all.

Chapter Seven. Although it is now well over thirty years old, the work that arguably still presents the most complete and balanced account of evolutionary principles is *The Major Features of Evolution* by the late George Gaylord Simpson (New York: Simon and Schuster, 1968, reprint of the 1953 edition). And the treatment of classification, also by George Gaylord Simpson, in *Principles of Animal Taxonomy* (New York: Columbia University Press, 1961), continues to be the best-rounded and most useful work on the subject.

The modern cladistic manifestation of the Medieval Aristotelian approach to the treatment of organic relationships derives principally from the works of the late Willi Hennig, especially his *Phylogenetic Systematics,* translated from the German by D. Dwight Davis and Rainer Zangerl (Urbana: Univeristy of Illinois Press, 1966). His principal tenets are perpetuated as a kind of unquestioned dogma in what could almost be regarded as a "cult" among his followers—for example, Steven M. Stanley in *Macroevolution: Pattern and Process* (San Francisco, Freeman, 1979), and *The New Evolutionary Timetable: Fossils, Genes, and the Origin of Species* (New York: Basic Books, 1981); in Niles Eldredge and Joel Cracraft's *Phylogenetic Patterns and the Evolutionary Process: Method and Theory in Comparative Biology* (New York: Columbia University Press, 1980); and in Edward O. Wiley's *Phylogenetics: The Theory and Practice of Phylogenetic Systematics* (New York: Wiley-Interscience, 1981).

The view that stresses punctuated equilibria can be regarded as a derived offshoot of cladistics, and it was initially proposed by Niles Eldredge and Stephen Jay Gould in "Speciation and Punctuated Equilibria: An Alternative to Phyletic Gradualism," in T. J. Schopf, ed., *Models in Paleobiology* (San Francisco: Freeman, 1972).

The molecular biological basis for the consequences of mutation in the absence of selection—in fact, the logic for the Probable Mutation Effect—was first noted in "Structural Reduction in Evolution," by C. L. Brace, *The American Naturalist,* Vol. 97, No. 1 (1963). A more extensive use of this viewpoint can be seen in *Evolution by Gene Duplication* by Susumu Ohno (New York: Springer-Verlag, 1970). More recently it has been discussed by Motoo Kimura, *The Neutral Theory of Molecular Evolution* (New York: Cambridge University Press, 1983); by Masatoshi Nei and Richard G. Koehn, eds., *Evolution of Genes and Proteins* (Sunderland, Mass.: Sinauer Associates, 1983), and by Tomoko Ohta and Kenichi Aoki, eds., *Population Genetics and Molecular Evolution* (New York: Springer-Verlag, 1985).

Chapter Eight. Modern efforts to appraise the overall picture of primate evolution stem largely from Elwyn L. Simons's *Primate Evolution: Introduction to Man's Place in Nature* (New York: Macmillan, 1971). A series of edited volumes—for instance, *The Functional and Evolutionary Biology of Primates* (Chicago: Aldine-Atherton, 1972) and *Primate Functional Morphology and Evolution* (The Hague: Mouton, 1973), Russell H. Tuttle, ed.; *Evolutionary Biology of the New World Monkeys and Continental Drift,* Russell L. Ciochon and A. Brunetto Chiarelli, eds. (New York: Plenum, 1980); *New Interpretations of Ape and Human Ancestry,* R. L. Ciochon and Robert S. Corruccini, eds. (New York: Plenum, 1983); and *Primate Evolution and Human Origins,* R. L. Ciochon and John G. Fleagle, eds. (Menlo Park, California: Benjamin/Cummings, 1985); together with *Evolutionary History of the Primates* by Frederick S. Szalay and Eric Delson (New York: Academic, 1979)—have provided extensive material that would otherwise be scattered through a great many technical and professional journals.

Chapter Nine. One of the earliest explicit realizations of the behavioral significance of hominid body form was the unappreciated paper by Paul Alsberg, "The Taungs Puzzle: A Biological Essay," *Man,* Vol. 34, No. 179 (1934).

The following four papers overlap to provide a coherent picture of specific hominid ecology, tool use, behavior, and cultural adaptation: G. A. Bartholomew and J. B. Birdsell, "Ecology and the Protohominids," *American Anthropologist,* Vol. 55, (No. 4 (1953); Marshall D. Sahlins, "The Origin of Society," *Scientific American,* Vol. 203, No. 3 (1960); S. L. Washburn, "Tools and Human Evolution," *Scientific American,* Vol. 203, No. 3 (1960); and C. L. Brace, "Biological Parameters and Pleistocene Hominid Life-Ways," in I. S. Bernstein and E. O. Smith, eds., *Primate Ecology and Human Origins: Ecological Influences of Social Organization* (New York: Garland Press, 1979).

For a review of the theory, and the techniques and practices of dealing with archaeological and geological time, see Stuart Fleming's *Dating in Archaeology: A Guide to Scientific Techniques* (New York: St. Martin's, 1977), and W. B. Harland, A. V. Cox, P. G. Llewellyn, C. A. G. Pickton, A. G. Smith, and R. Walters's, *A Geologic Time Scale* (New York: Cambridge University Press, 1983), which provide the necessary background.

Chapter Ten. The initial modest paper by R. A. Dart, remarkable for the furor it touched off, is still worth reading: "*Australopithecus africanus:* The Man-Ape of South Africa," *Nature,* Vol. 115 (February 7, 1925). The dental and postcranial evidence for the South African Australopithecines was treated by John T. Robinson in *The Dentition of the Australopithecinae* (Pretoria: Transvaal Museum Memoir No. 9, 1956), and *Early Hominid Posture and Locomotion* (Chicago: University of Chicago Press, 1972).

The appraisal of the first major Australopithecine discovery in East Africa was published by Phillip V. Tobias, *The Cranium and Maxillary Dentition of Australopithecus (Zinjanthropus) Boisei,* Vol. 2 of *Olduvai Gorge,* L. S. B. Leakey, ed. (Cambridge, Eng.: Cambridge University Press, 1967). An interim survey of the East Turkana discoveries can be found in Meave G. Leakey and Richard E. Leakey, eds., *Koobi Fora Research Project. Volume I, The Fossil Hominids and an Introduction to their Context, 1968–1974* (Oxford: Clarendon Press, 1978). Using the Hadar material as the basis for a somewhat different interpretation, Donald C. Johanson and Tim D. White presented "A Systematic Assessment of Early African Hominids. Primates: Hominidae, *Science,* Vol. 203 (1979). Further exemplary reports on the Ethiopian Australopithecines can be found in: D. C. Johanson, T. D. White, and Y. Coppens, "Dental remains from the Hadar Formation, Ethiopia: 1974–1977 Collections," *American Journal of Physical Anthropology,* Vol. 57, No. 4 (1982); D. C. Johanson, C. O. Lovejoy, W H. Kimbel, T. D. White, S. C. Ward, M. E. Bush, B. M. Latimer, and Y. Coppens, "Morphology of the Pliocene partial hominid skeleton, A. L. 288–1 from the Hadar Formation, Ethiopia," *American Journal of Physical Anthropology,* Vol. 57, No. 4 (1982); D. C. Johanson, M. Taieb, and Y. Coppens. "Pliocene Hominids from the Hadar Formation, Ethiopia (1973–1977): Stratigraphic, Chronologic, and Paleoenvironmental Contexts with Notes on Hominid Morphology and Systematics," *American Journal of Physical Anthropology,* Vol. 57, No. 4 (1982); William H. Kimbel, T. D. White, and D. C. Johanson, "Craniodental Morphology of the Hominids from Hadar and Laetoli: Evidence of '*Paranthropus*' and *Homo* in the Mid-Pliocene of Eastern Africa?" in *Ancestors: The Hard Evidence,* Eric Delson, ed. (New York: Alan Liss, 1985). Another survey of Australopithecine craniofacial form can be found in Yoel Rak's *The Australopithecine Face* (New York: Academic Press, 1983).

Finally, there have been some concerted efforts to analyze the evidence concerning just what the Australopithecines were doing for a living, and much of this has been published in the volume edited by Juliet Clutton-Brock and Caroline Grigson, *Animals and Archaeology: 1. Hunters and their Prey* (Oxford: British Archaeological Reports, BAR International Series 163, 1983). In this are found papers by the late Glynn Isaac, "Bones in Contention: Competing Explanations for the Juxtaposition of Early Pleistocene Artifacts and Faunal Remains"; by Henry T. Bunn, "Evidence of the Diet and Subsistence Patterns of Plio-Pleistocene Hominids at Koobi Fora, Kenya, and Olduvai Gorge, Tanzania"; and by Pat Shipman, "Early Hominid Lifestyle: Hunting and Gathering or Foraging and Scavenging." The most recent assessment of this problem is by Pat Shipman, "Scavenging or Hunting in Early Hominids: Theoretical Framework and Tests," *American Anthropologist,* Vol. 88, No. 1 (1986).

Chapter Eleven. The standard reference against which all accounts of the Pithecanthropines are compared is Franz Weidenreich's "The Skull of *Sinanthropus pekinensis," Palaeontologia Sinica,* Vol. 10 (1943). Presenting a more historically oriented picture is G. H. R. von Koenigswald's "The Discovery of Early Man in Java and Southern China," in W. W. Howells, ed., *Early Man in the Far East,* Studies in Physical Anthropology, No. 1 (American Association of Physical Anthropologists, 1949).

More recently, a summary treatment has been offered in K. W. Butzer, G. L. Isaac, E. Butzer, and B. Isaac, eds., *After the Australopithecines: Stratigraphy, Ecology, and Culture Change in the Middle Pleistocene* (The Hague: Mouton, 1975). The extensive postwar Javanese finds, as well as earlier discoveries, are surveyed by Teuku Jacob, "The Pithecanthropines of Indonesia," *Bulletins et Mémoires de la Société d'Anthropologie de Paris, 2, série* 13 (1975). A systematic attempt to deal with the dating problems in Java has been undertaken by Shuji Matsu'ura, "A Chronological Framing for the Sangiran Hominids: Fundamental Study by the Fluorine Dating Method," *Bulletin of the National Science Museum, Tokyo,* Series D (Anthropology), Vol 8 (1982).

The evidence for actual skeletal material of Pithecanthropine form in India, to go along with the abundant archaeological material, is reported by Arun Sonakia, "Early *Homo* from Narmada Valley, India," in Eric Delson, ed., *Ancestors: The Hard Evidence* (New York: Alan Liss, 1985).

One of the problems in dealing with the relationships between *Homo erectus* and *Homo sapiens* is that those of a cladistic bent, who will not allow the view that one species could gradually become transformed into another one, will redefine what is obviously at the *erectus* or Pithecanthropine Stage and call it *"sapiens,"* even though it has all the adaptive features of a good Pithecanthropine. This is what is done, for example, by Chris B. Stringer in "The Definition of *Homo erectus* and the Existence of the Species in Africa and Europe," in P. Andrews and J. L. Franzen, eds., *The Early Evolution of Man* (Senckenberg: Courier Forschungsinstitut, 1984). A similar preference is also exhibited by Günter Bräuer in "The 'Afro-European *sapiens*–Hypothesis,' and Hominid Evolution in East Asia During the Late Middle and Upper Pleistocene," in P. Andrews and J. L. Franzen, eds., *The Early Evolution of Man: With Special Emphasis on Southeast Asia and Africa* (Senckenberg: Courier Forschungsinstitut, 1984). The preliminary information on the West Turkana specimen, WT 15,000, is available in the paper by Frank Brown, John Harris, Richard Leakey, and Alan Walker, "Early *Homo erectus* Skeleton from West Lake Turkana, Kenya," *Nature,* Vol. 316 (1985). Further information has been provided by Alan Walker, *"Homo erectus* Skeleton from West Lake Turkana, Kenya," *American Journal of Physical Anthropology,* Vol. 69, No. 2 (1986).

Chapter Twelve. The definitive treatment of the morphology of the Kabwe or "Rhodesian" cranium as well as that of Petralona is by Rupert I. Murrill, *Petralona Man: A Descriptive and Comparative Study, With New Important Information on Rhodesian Man* (Springfield, Illinois: C. C. Thomas, 1981). Saldanha has been appraised by M. R. Drennan, "The Special Features and Status of the Saldanha Skull," *American Journal of Physical Anthropology,* Vol. 13, No. 4 (1955).

The classic study of the Solo material was done by Franz Weidenreich in his "Morphology of Solo Man," *Anthropological Papers of the American Museum of Natural History,* Vol. 43, Part 3 (1951). A more recent quantitative treatment that does not include any perspective on evolutionary biology is by Albert P. Santa Luca, "The Ngandong Fossil Hominids," *Yale University Publications in Anthropology,* Vol. 78 (1980). A similarly limited treatment of the Petralona cranium was offered by C. B. Stringer, "A Multivariate Study of the Petralona Skull," *Journal of Human Evolution,* Vol. 3 (1974).

The Dali skull has been the subject of excellent and thoughtful studies by Wu Xinzhi in "A Well-Preserved Cranium of an Archaic Type of *Homo sapiens* from Dali, China," *Scientia Sinica,* Vol. 24, No. 4 (1981), and by Wu Xinzhi and Wu Maolin in "Early *Homo sapiens* in China," in Wu Rukang and John W. Olsen, eds., *Palaeoanthropology and Palaeolithic Archaeology in the People's Republic of China* (Orlando: Academic Press, 1985).

Swanscombe and Fontéchevade were splendidly treated by J. S. Weiner and B. G. Campbell in "The Taxonomic Status of the Swanscombe Skull," in C. D. Ovey, ed., *The Swanscombe Skull* (London: The Royal Anthropological Institute, 1964). The most recent appraisal of the Steinheim cranium is by Karl D. Adam, "The Chronological and Systematic Position of the Steinheim Skull," in Eric Delson, ed., *Ancestors: The Hard Evidence* (New York: Alan Liss, 1985).

Arguments concerning the vexing issue of assessing the evidence for the control of fire have gone from extravagant claims to extreme skepticism. Prominent among the skeptics, especially where the issue is the evidence for the control of fire at Zhoukoudian, are Lewis R. Binford and Chuan Kun Ho, "Taphonomy at a Distance: Zhoukoudian, 'The Cave Home of Beijing Man'?" *Current Anthropology,* Vol. 24, No. 4 (1985).

Chapter Thirteen. Many, if not most, physical anthropologists prefer the approach to the Neanderthal problem represented in the writings of F. Clark Howell—for instance, in "The Place of Neanderthal Man in Human Evolution," *American Journal of Physical Anthropology,* Vol. 9, No. 4 (1951); and "The Evolutionary Significance of Variation and Varieties of 'Neanderthal' Man," *Quarterly Review of Biology,* Vol. 32, No. 4 (1957). For a collection of papers generally in the traditional frame of reference, see G. H. R. von Koenigswald, ed., *Hundert Jahre Neanderthaler: Neanderthal Centenary* (Utrecht: Kemink en Zoon, 1959).

A rather different view, obviously more in line with the approach taken in this book, is presented by C. L. Brace in "Refocussing on the Neanderthal Problem," *American Anthropologist,* Vol. 64, No. 4 (1962); "Krapina, 'Classic' Neanderthals, and the Evolution of the European Face," *Journal of Human Evolution,* Vol. 8, No. 5 (1979); and in "Tales of the Phylogenetic Woods: The Evolution and Significance of Evolutionary Trees," *American Journal of Physical Anthropology,* Vol. 56, No. 4 (1981). More in line with this approach is the stance taken by Fred H. Smith and Gail C. Ranyard, "Evolution of the Supraorbital Region in Upper Pleistocene Fossil Hominids from South Central Europe," *American Journal of Physical Anthropology,* Vol. 53, No. 4 (1980); Fred H. Smith, "Upper Pleistocene Hominid Evolution in South-Central Europe: A Review of the Evidence and Analysis of Trends," *Current Anthropology,* Vol. 23, No. 6 (1982); and Frank Spencer, "The Neandertals and their Evolutionary Significance: A Brief Historical Survey," in F. H. Smith and F. Spencer, eds., *The Origins of Modern Humans* (New York: Alan Liss, 1984).

Old views die hard, however, as is shown in a series of papers that demonstrate the continuity of the dominant orthodoxy that would deny the fossil evidence for human evolution: W. W. Howells, "Neanderthals: Names, Hypotheses, and Scientific Method," *American Anthropologist,* Vol. 86, No. 1 (1974); "Neanderthal Man: Facts and Figures," *Yearbook of Physical Anthropology, 1974,* Vol. 18 (1976); Bernard Vandermeersch, "Les Hommes Modernes et les Néandertaliens Ont Cohabité: Le Néandertalien de Saint'Césaire," *La Recherche,* Vol. 119, No. 12 (1981); "The Origin of the Neandertals," in Eric Delson, ed., *Ancestors: The Hard Evidence* (New York: Alan Liss, 1985); C. B. Stringer, "Towards a Solution to the Neanderthal Problem," *Journal of Human Evolution,* Vol. 11, No. 5 (1982); "Fate of the Neanderthal," *Natural History,* Vol. 93, No. 12 (1984); "Middle Pleistocene Hominid Variability and the Origin of Late Pleistocene Humans," in Eric Delson, ed., *Ancestors: The Hard Evidence* (New York: Alan Liss, 1985).

The suggestion that Neanderthals had a year-long gestation period was initially made by Erik Trinkaus, "Neanderthal Pubic Morphology and Gestation Length," *Current Anthropology,* Vol. 25, No. 4 (1984). This has been convincingly refuted by Karen R. Rosenberg, *The Functional Significance of Neanderthal Pubic Morphology* (Ann Arbor, Michigan: University Microfilms, 1986).

Chapter Fourteen. The fictional portrayal of old-fashioned hominid catastrophism in modern "undress" is represented by Jean M. Auel's *The Clan of the Cave Bear: Earth's Children* (New York: Crown Publishers, 1980), and its best-selling sequels. A similar but more informed and much better written book is *Dance of the Tiger: A Novel of the Ice Age* by Björn Kurtén (New York: Pantheon, 1980). The "intellectual" basis for their expectations, such as it is, can be found in perfunctory form in Niles Eldredge and Ian Tattersall's *The Myths of Human Evolution* (New York: Columbia University Press, 1982).

The transition from Neanderthal to Modern form is best represented by the skeletal remains from Skhūl at Mount Carmel, Israel. The classic description was published by Theodore D. McCown and Sir Arthur Keith in *The Stone Age of Mount Carmel: The Fossil Human Remains from the Levalloiso-Mousterian* (Oxford: Clarendon Press, 1939). The similar and simultaneous change taking place further west—in east and southeast Europe—is documented by Fred H. Smith, "Fossil Hominids from the Upper Pleistocene of Central Europe and the origin of Modern Europeans," in F. H. Smith and F. Spencer, eds., *The Origins of Modern Humans* (New York: Alan Liss, 1984), and "Continuity and Change in the Origin of Modern *Homo sapiens,*" *Zeitschrift für Morpoholgie und Anthropologie,* Vol. 75, No. 2 (1985).

The treatment of the origin of modern human form in the volume *The Origin of Modern Humans: A World Survey of the Fossil Evidence,* Fred H. Smith and Frank Spencer, eds.

(New York: Alan Liss, 1984), is only partially satisfactory at best. This is because of the continuing presence of old-fashioned, antievolutionary hominid catastrophism in essays such as Günter Bräuer's "A Craniological Approach to the Origin of Anatomically Modern *Homo sapiens* in Africa and Implications for the Appearance of Modern Europeans," and C. B. Stringer, J. J. Hublin, and B. Vandermeersch's "The Origin of Anatomically Modern Humans in Western Europe."

The evidence that human evolution has continued and even accelerated since the end of the Pleistocene is documented by David W. Frayer, *Evolution of the Dentition in Upper Paleolithic and Mesolithic Europe* (Lawrence: University of Kansas Publications in Anthropology, No. 10, 1978); by C. L. Brace, "Australian Tooth-Size Clines and the Death of a Stereotype," *Current Anthropology,* Vol. 21, No. 2 (1980); by C. L. Brace and Virginia Vitzthum, "Human Tooth Size at Mesolithic, Neolithic and Modern Levels at Niah Cave, Sarawak: Comparisons with Other Asian populations," *Sarawak Museum Journal,* Vol. 30, No. 54 (1984); and by C. L. Brace, Karen R. Rosenberg, and Kevin D. Hunt, "Late Pleistocene and Post-Pleistocene Change in Human Tooth Size: A Case of Evolutionary Gradualism," *Evolution,* in press.

The single best available work that covers the whole spectrum of human evolution and which tries to keep a Darwinian perspective on it all is *Paleoanthropology* by Milford H. Wolpoff (New York: A. A. Knopf, 1980). A compatible, if somewhat dated, visual supplement can be provided by the *Atlas of Human Evolution* by C. L. Brace, Harry Nelson, Noel Korn, and Mary L. Brace (New York: Holt, Rinehart & Winston, 1979).

Index